Getting started with Arduino and Raspberry Pi

Christopher Dock

Copyright © 2021 Christopher Dock

All rights reserved.

ISBN: 978-1-952930-02-7

No portion of this book may be reproduced in any form without permission from the publisher, except as permitted by U.S. copyright law. For information about special discounts available for bulk purchases, sales promotions, fund-raising and educational needs contact us.

North American Offices
Christopher Dock
3533 Rum River Drive
Anoka, MN 55303
USA

European Offices
Christopher Dock
Ludwig-Richter-Straße 15
60433 Frankfurt
Germany

Limits of Liability and Disclaimer of warranty
We work hard to provide the most accurate research on options for self-publishing books. All information provided in Getting started with Arduino and Raspberry Pi has been provided in good faith, however, we make no representations, warranty or guarantees of any kind, express or implied, that this information will be complete, accurate, adequate, error free or will not change in the future. Information in this book may not constitute the most up-to-date information available. Getting started with Arduino and Raspberry Pi is for general informational purposes only and should not be relied upon in making legal or financial decisions.

To the maximum extent permitted by law, we disclaim all warranties, whether express or implied, including the implied warranties of merchantability and fitness for a particular purpose. To the maximum extent permitted by law, we disclaim any and all liability in the event any information, commentary, analysis, or opinions contained in this book prove to be inaccurate, incomplete or unreliable, and under no circumstances will we be liable to you for any lost profits, revenues, information, or data or incidental damages arising out of or related the information provided by this book.

Information provided in this book is descriptions about electronics as well as various tools that can be used to make your own electronic creations. The tips and information is intended to be helpful but should not be considered a replacement for any manuals, documentation or training for operation of such tools.

DEDICATION

Mikhael, thanks for introducing me to a whole new world of microelectronics. It was neat learning the tools and techniques from someone whose history extends from pre-circuit boards up until circuit boards created using surface mount technology. It was invaluable learning at the knee of a master; the ability to create and enhance your own electronics or to program them to create my own IOT devices is both liberating and satisfying.

CONTENTS

	Preface	1
	Introduction	4
1	Theory and components	13
2	Tools	42
3	Simple circuits	64
4	Integrated circuits	74
5	Software tools	98
6	Arduino	108
7	Raspberry Pi	123
8	Streaming audio	153
9	Network storage	165
10	Build LED Arduino cube	172
11	LED cube with custom PCB	186
12	"C" programming refresher	202
13	Makefiles	222
14	Kicad	230
15	Linux commands	246
16	Advance Linux commands	251
17	SMD size chart	262
	Appendix Arduino Cube pattern code	263
	Glossary	270

ACKNOWLEDGMENTS

I cannot offer enough heartfelt thanks to Carsten Wenzke for all his tireless efforts. He has not only read every word to ensure what I have written is understandable but has gone much further. He has searched for inconsistencies, shortcomings or downright logic gaps and offered honest suggestions designed to bring about the best final product.

Index of Tables

Table 1: *Personal computer specs over the years*..8

Table 2: *Original Raspberry Pi specs*..9

Table 3: *Raspberry Pi 4 specs*...10

Table 4: *Various batteries and their attributes*..20

Table 5: *Wire dimensions in imperial and metric sizes*...22

Table 6: *Solid wire*...23

Table 7: *Stranded wire*..23

Table 8: *Through hole resistor color table*..29

Table 9: *Different types of solders and their attributes*...55

Table 10: *Parts list for train flasher circuit*...66

Table 11: *Segment setup for 7 segment display*...85

Table 12: *Binary coded decimal values*..85

Table 13: *IOCON.BANK = 0*...89

Table 14: *Parts list for 555 flasher*...93

Table 15: *Parts list for micro-controller driven flasher*...96

Table 16: *Modified parts for train flasher circuit*..97

Table 17: *Example flow for transferring 3 bytes using I2C*..119

Table 18: *Samba share configuration*..168

Index of Illustrations

Illustration 1: Sample schematic..13

Illustration 2: Electron flow in a circuit..15

Illustration 3: Graph of direct voltage over time..16

Illustration 4: Single phase alternating current over time..17

Illustration 5: Three phase alternating current...18

Illustration 6: Circuit that is only power to drive LED...21

Illustration 7: Breadboard..24

Illustration 8: Single sided copper prototype board..26

Illustration 9: Prototype board with plated through holes..27

Illustration 10: Through hole resistors color example...29

Illustration 11: Example of capacitor having leaked onto the circuit board...................................35

Illustration 12: Batteries in series..36

Illustration 13: Batteries in parallel...37

Illustration 14: Incorrect schematic..41

Illustration 15: Safety glasses..42

Illustration 16: Head mounted magnifier..42

Illustration 17: Fume extractor..43

Illustration 18: Medical mask..43

Illustration 19: Regular multimeter..44

Illustration 20: High end multimeter..44

Illustration 21: Flushcut wire cutter..47

Illustration 22: Best tweezers..47

Illustration 23: Lower quality...47

Illustration 24: Board holder...48

Illustration 25: Helping hand..48

Illustration 26: Piece of circuit board..49

Illustration 27: Homemade breakout cable...49

Illustration 28: Mosfet tester cable...50

Illustration 29: Wire with LED and inline resistor...50

Illustration 30: 9V battery connector with inline resistor..51

Illustration 31: Pen style soldering iron..51

Illustration 32: Soldering station...52

Illustration 33: Multiple styles of soldering tips..52

Illustration 34: Soldering iron holder...53

Illustration 35: Example of a ruined part..57

Illustration 36: Situation where heat transfer can create soldering problems..................................58

Illustration 37: Step 1, Place wire on the solder joint..59

Illustration 38: Step 2, Press wire into solder joint...59

Illustration 39: De-soldering wick...59

Illustration 40: Solder sucker...60

Illustration 41: Manual stoplight schematic..65

Illustration 42: Manual stoplight breadboard test..65

Illustration 43: Flasher circuit schematic...66

Illustration 44: Flasher circuit breadboard test...66

Illustration 45: Incorrect circuit with pull up resistor and button...67

Illustration 46: Circuit with pull up resistor and button...67

Illustration 47: Alternating current graphed over time..71

Illustration 48: Output from half wave rectifier..71

Illustration 49: Half wave rectifier..71

Illustration 50: Full rectifier...72

Illustration 51: Output from full wave rectifier...72

Illustration 52: Input on left side of bridge positive...72

Illustration 53: Input on left side of bridge negative..72

Illustration 54: 555 Timer chip..75

Illustration 55: 555 timer chip layout...75

Illustration 56: Monostable circuit over time..76

Illustration 57: Monostable circuit schematic...76

Illustration 58: Astable circuit over time..77

Illustration 59: Astable circuit schematic...78

Illustration 60: Bistable circuit schematic..79

Illustration 61: Micro-controller functionality..80

Illustration 62: Fuse calculator..81

Illustration 63: Fuse calculator continued..81

Illustration 64: Fuse calculator continued..82

Illustration 65: I/O Expander...87

Illustration 66: Modified toy ambulance..90

Illustration 67: A stable circuit using 555 timer..90

Illustration 68: Modified astable circuit schematic with alternate flashing.....................................91

Illustration 69: Location for flashing lights..92

Illustration 70: Ambulance front end assembly...92

Illustration 71: Flashing circuit..92

Illustration 72: Micro-controller driver flasher circuit schematic...93

Illustration 73: Breadboard implementation of micro-controller driven circuit............................94

Illustration 74: 10 pin connector layout of ISP connector...95

Illustration 75: Sample schematic in schematic editor...99

Illustration 76: PCB layout of an astable 555 circuit..100

Illustration 77: Breadboard view of circuit schematic in fritzing editor...101

Illustration 78: AVR Dragon with ZIF socket..104

Illustration 79: USBTinyISP...105

Illustration 80: Arduino integrated development environment with open sketch.....................106

Illustration 81: Micro controller device with ISP programmer...111

Illustration 82: Arduino connected with USB cable...112

Illustration 83: Pulse width modulation showing several different duty cycles..........................116

Illustration 84: Windows SD card image burner..128

Illustration 85: Raspi-config program...129

Illustration 86: First boot dialog..131

Illustration 87: Putty configuration..132

Illustration 88: Raspberry Pi GUI for setting up system daemons...134

Illustration 89: Start WiFi configuration..136

Illustration 90: Step #1..137

Illustration 91: Step #2 & Step 3..137

Illustration 92: Step #4..138

Illustration 93: Step #5, final setup..138

Illustration 94: Completed 3x3x3 cube Arduino shield..173

Illustration 95: Controlling a single LED from a LED grid...174

Illustration 96: 3x3x3 cube schematic controlled by Arduino..176

Illustration 97: Led cube showing pin mapping..177

Illustration 98: ELEGOO ATmega2560..177

Illustration 99: PCB for Arduino MEGA 2560...178

Illustration 100: A form for creating 3x3x3 or 4x4x4 layer...178

Illustration 101: Cube pins soldered to the Arduino pins...180

Illustration 102: Showing the names of each column..180

Illustration 103: First self designed - front...187

Illustration 104: First self designed - back...187

Illustration 105: Chip powering LED...188

Illustration 106: Chip controlling power to LED...188

Illustration 107: 3x3x3 cube schematic..189

Illustration 108: Proper schematic for the column..191

Illustration 109: 3x3x3 attempts at control boards..192

Illustration 110: 3x3x3 printed circuit board..193

Illustration 111: 3x3x3 board with all parts soldered to it...196

Illustration 112: Partial layer..197

Illustration 113: Full layer...197

Illustration 114: Kicad main dialog..230

Illustration 115: Schematic with all wires connected..231

Illustration 116: Schematic with labels showing relationships..232

Illustration 117: Assign footprints dialog..234

Illustration 118: Side view of pcb...236

Illustration 119: Surface mount part connecting to a through hole..236

Illustration 120: Example of a via...237

Illustration 121: Star configuration..238

Illustration 122: Daisy chained configuration..238

PREFACE

I love computers and as a software developers I can hardly imagine living in a world where you cannot use them to solve all your little problems. When you need to generate some forms or search through data for some pattern you write a program. This has been made all that much easier due to the great quantity of freely available open source software. I really didn't think that it could get any better and then I discovered electronics. There is nothing that is more exciting than creating your own electronic device. Create the schematic, design the board, attach the parts and then write a program to control it. You turn on your device and know that you created every part of it from the ground up.

With this particular revelation I decided to create a small guide that would provide a starting point so my boys could also experience the excitement of creating their own projects as well as to use computers for something other than just playing either video games.

This "guide" eventually morphed into something a bit larger as well as providing a more comprehensive background. The goal is to have a single source which can be used to provide a proper start to anyone who is interested in creating their own electronic or IoT device or get a start using open source software.

This book is intended to help the reader to learn by hands on work but also provide reference material.

Navigating through the book
The content in this book doesn't need to be read in any particular order, you can skip over sections you are already familiar with. This book can essentially be broken into two different parts. The first half is a description of the individual theory, components, tools, software and micro platforms that can be used to build your projects. The second half is a few projects that are more in-depth and to showcase the tools and provide some examples.

Below is a summary of what each chapter will cover. They can essentially be read in any order but the later chapters do assume some knowledge from the earlier chapters.

    Building blocks
        Chapter 1 – Introduction
        Chapter 2 – Theory and components
        Chapter 3 – Tools
        Chapter 4 – Simple circuits

Chapter 5 – Integrated circuits
Chapter 6 – Software Tools

Platforms
Chapter 7 – Arduino
Chapter 8 – Raspberry Pi

Projects
Chapter 9 – Streaming audio
Chapter 10 – Network storage
Chapter 11 – build LED cube using Arduino
Chapter 12 – build LED cube with custom PCB

Reference Data
Appendix – "C" programming refresher
Appendix – Kicad
Appendix – Linux commands
Appendix – Advanced Linux commands
Appendix - Makefiles
Glossary

Assumptions

This book covers a lot of different areas starting with basic electronic components and micro-controllers and working through software development and circuit board design. The assumption is a low to no experience with electronics or small micro electronics but perhaps some experience with computers and software development.

This book is starting from a beginners level and working through to creating and programming our own electronic device. Enough information is provided that a novice can get started with any of the topics in this book. This will allow you to become familiar with Arduino and Raspberry Pi and manage to use them in the scope of a larger DIY project. With some extra effort it is possible to go beyond these micro processor platforms and create your own device.

Despite this book telling a narrative from resistor through to completed device it is not a fantasy novel you would read while laying at the beach. This book can be used as a guide and as a reference but despite all the information covered it cannot answer every question sometimes more information is necessary.

This information might be found in other books but more and more often it is found on the internet, provided by parts vendors or even in the form of a video on Youtube or other hobbyist site.

Font Conventions

The following conventions have been used in this book.

`constant width`	Code examples, fragments, as well as other text related to the code or the building thereof.
bold	Is used in examples to demonstrate commands that must be typed exactly.
italic	Is used for displaying the output from commands or programs.

$	Is the prompt of the Bash shell where commands are typed.
#	Is the prompt from the bash shell where the root user is logged in.
c:\	Is the windows command prompt from where commands are typed.

INTRODUCTION

DIY - do it yourself.

For years computers have been getting cheaper and faster all the while the quantity and quality open source software have been meeting and exceeding quality of propitiatory solutions. It is now possible to setup an application server, database server or a web server running on a Linux machine without spending a dime on software. Set all of this up and connect the machine to the internet and you could have your own web site or perhaps even a small e-commerce site.

This is really neat! The entire Linux ecosystem makes it possible to create solutions. You can get programming languages, compilers, utilities, create scripts or schedule batch processes. There is really almost nothing you cannot do.

This isn't 100% correct. Personal computers have a lot of power but they are not very small. In general, the size a standard desktop personal computer is too large and too expensive to include into small projects around the house. For the longest time, I had been looking for a computer that would be both small enough to create my own smart device as well as cheap enough I wouldn't have to worry about the cost.

Good news is that a lot of different smart sockets, embedded computers or micro-controller boards have hit the market and this is no longer a problem. Due to the amazing advances in technology through miniaturization and cost reduction it is possible to get quite small Linux based computers and micro-controller boards which can be programmed for any special tasks. Over the last decade both the power of these family of devices has increased as well as the ever growing ecosystem of sensors and special purpose boards to be attached to them.

One of the most important differences between these micro-controller devices and personal computers is the ability to easily interact with the world around us. This isn't to say that you cannot interact using computers, they have PCI slots for cards, serial ports and USB ports but none of these are an especially easy way to interface with a computer. Small embedded devices are perfect for adding to intelligence to a typical home project and have special pins that are easily accessible and controllable.

It is no longer necessary to wait for Megacorp to create the next Bluetooth speaker or smart appliance – we can do it ourselves. It is an empowering time not only due to the advances in electronics but also the amazing quality and quantity of the tools.

- 3D Printers
- CNC Milling
- Drill presses
- Printed Circuit Board manufactures
- Cloud Services
- Power tools (e.g. electric drill or circular saw)
- High speed internet

Just using these tools and services alone would help to outfit a very nice research and development department but this is only just a few of the available resources.

Other resources.

Software	Open Source(free)	Closed source
Desktop publishing	Scribus	Adobe InDesign[1]
Photo editing	Gimp	Adobe Creative Cloud[2]
Word processing	Libre Office	Open Office 365[3]
Operating System	Linux, BSD	Windows
Web server	Apache	Internet Information Service
Application server	Tomcat	Internet Information Service
Blogging software	Wordpress	Wix
Printed circuit board CAD	Kicad	Eagle, PROTEL
Email client	Thunderbird	Microsoft Outlook
Email server	Postfix	Microsoft Exchange Server
Testing	Postman	Uptrends
Load Testing	Gatling, Jmeter	LoadRunner
Integrated development environment	Eclipse	Microsoft Visual Server, IntelliJ[4]
Entertainment, Modeling	Blender	Apple Motion

Hardware	Open Source(free)	Closed source
Micro-controller platform	Arduino	Beaglebone
Computer	Raspberry Pi	

These tools can be used to create a home entertainment system, run a small business (or a large one) or just create some fun projects.

As much as I would like to cover all programs that might be interesting in DIY projects I will be limited describing information about the hardware platforms from how to set it up to program it. In some cases installing additional software to get the most from it.

I would like to document as many of the useful, in most cases free, tools and how to use them. I would also like to give an in depth how to of these same tools, lessons learned and a lot of undocumented tricks to speed things up. To do this for all of the tools listed earlier as well as embedded computing and micro controllers is unfortunately unrealistic.

---

1 Monthly subscription

2 Monthly subscription

3 Monthly subscription

4 IntelliJ also has a free community edition

I am going to cover as completely as possible enough information that it is possible to select the hardware to be embedded in your project, see how to connect and or program then as well as show the basics in order to create your own schematic and get a PCB manufactured from it.

This will include some examples and clarifications for using or even creating your own electronics in order to add intelligence to your projects. While fairly comprehensive, the material in this book is intended to assist you, the reader, to get started and to focus on and further investigate topics that you find interesting.

This book does assume some familiarity with computers and software development as well as the interest in electronics. Although there will be some reference material to assist it is beyond the scope of this book to make you an expert in Linux, software development or electronics. This book's goal is to provide enough information to get started for people with the desire to explore and learn other technologies.

Arduino

The Arduino[5] is probably the most recognized open source hardware project to date. It is not just the software that controls the device is available but also the actual schematic for the device itself. You can download the manual, schematic, and even eagle files to examine or even to extend it yourself.

Just like many other saga's, the story of the Arduino started a long long time ago, in 2005, which for this story was taking place in Ivrea Italy. All the conditions which helped to create the need for a small affordable micro-controller on a board were all coming together. Like many other inventions, the Arduino came into being because there was a need that was not being fulfilled by the market. The difference in this case was the solution was not fulfilled by a large corporation with hundreds of developers but actually based on an idea from a professor at the Interaction design institute in Ivrea Italy.

Associate professor Banzi was teaching about novel ways of interactive design which was essentially physical computing. The Banzi as well as the other professors were using the BASIC Stamp micro-controller developed by Parallax. The students had been making use of this micro-controller for at least a decade. The Stamp was programmed using the BASIC programming language. The platform was a small circuit board, memory, micro-controller and of course input and output ports. It was pretty much what you would expect for a micro-controller board.

However, the students encountered the fairly common constraints often seen in the IT world – price and computing power. Cost is a relative and students usually have a much lower tolerance to costs and the $100 price tag of the BASIC Stamp was an issue. Not only was it relatively expensive but due to the speed it was not always computationally possible for the students to implement their ideas.

It was also about this time that a new programming language called "Processing" from MIT which also had a user friendly integrated development environment. It was the intersection of all of these facts plus a development from one of Banzi's students, Hernando Barragan. Hernando created a prototyping platform called Wiring which included both a IDE and a ready to use circuit board.

---

5 http://Arduino.cc

All of this gave Banzi the idea of creating a cheaper and easy to use platform. The prototype of which was first released in 2005 and later renamed to the familiar name Arduino.

Not a bug, it's a feature
One of the more unique ideas was to make the Arduino freely available much like open source software. However, open source software is covered by copyright which doesn't actually extend to cover hardware.

The solution to this problem was to use the creative commons "Attribution-ShareAlike" license. There are a number of different variations but in general this license makes it possible to share and adapt from an original without getting permission from the original owner. Not only is it possible to use the original you can do this for free, well with one or two small caveats.

Attribution — You must give appropriate credit, provide a link to the license, and indicate if changes were made. You may do so in any reasonable manner, but not in any way that suggests the licensor endorses you or your use.

ShareAlike — If you remix, transform, or build upon the material, you must distribute your contributions under the same license as the original.

https://creativecommons.org/licenses/by-sa/3.0/

Thus using the creative commons license and making the schematic publicly available an entire ecosystem came to life.

It was this licensing that helped to propel the Arduino from relative obscurity to perhaps most well known micro-controller board for the do-it-yourselfer crowd. Anyone who has a talent with or access to the proper tools could create their own Arduino or even an improved clone.

It was because of the uniquely affordable and friendly environment normal that helped to attract people with their own personal project ideas. The number of projects and number of devices that can be used by the Arduino or its compatible devices was consistently seen to increase. Not all the small electronic devices and sensors were in the format that could easily be used with the Arduino which then LED to the rise of the shield. An Arduino shield is a small circuit board that can be attached directly on top of the Arduino to the header pins.

One of the special features of the Arduino was the designers made a special effort that all the pins from the Atmel micro-controller powering the Arduino were accessible. It is this access to the micro-controller pins that allows anyone with a small amount of technical ability to connect up the Arduino to other external sensors using common components such as wire, resistors, capacitors, and LED's just to name a few. This forward thinking doesn't limit how the Arduino can be used or force the designers world view on the users of the Arduino.

Another feature that may have helped to sell the idea of the Arduino versus any other solutions at that time is that all the tools necessary for programming it are available for free.

Here is just a very tiny number of interesting projects that have been done using the

Arduino over time.

> RFID cat operated door
> http://duino4projects.com/rfid-cat-door-using-Arduino/
>
> LED watch
> http://duino4projects.com/led-watch-using-an-Arduino/
>
> Obstacle avoiding robot
> https://www.electronicshub.org/obstacle-avoiding-robot-Arduino/
>
> Wireless soil water monitor
> https://create.Arduino.cc/projecthub/55369/wireless-soil-moisture-probe-with-helium-and-dfrobot-c620b9?ref=tag&ref_id=sensor&offset=24

This is not to suggest that the Arduino was the only player in this particular space. They have shown considerable staying power progress considering that over the years some very big players have waded into the market.

Not quite a computer
The Arduino was able to keep its costs down in part because the goal wasn't to extract the maximum amount of profit per board but also because it wasn't a full fledged general purpose computer.

Specifications	Year
Personal computer Intel I5 quad core computer 4 gigabytes ram, 1 terabyte hard disk, 2.8Ghz, 64 bit	2017
IBM PC Intel 8088 computers 256kb ram, 10 mb hard disk, 4.77 mhz, 8 bit	1983
Apple II 6502 computer 48 kb ram, 143 kb floppy disk, 1 mhz, 8 bit	1977

Table 1: *Personal computer specs over the years*

The specifications of the Atmel micro-controllers don't compare well against the current IBM compatible systems but when compared to some of the original Apple II computers they look pretty impressive.

> **Specifications**
> Atmega328 micro-controller, 32 kb ram, 16 mhz, 8 bit
>
> Atmega1284 micro-controller, 128 kb ram, 20 mhz, 8 bit
>
> Atmega2560 micro-controller, 256 kb ram, 16 mhz, 8 bit

You can see that the lines begin to blur between what was originally a computer versus what today is just a small chip to assist or do dedicated tasks.

Perhaps that isn't a fair comparison as the Apple II is a proper computer with a proper operating system. That is both the advantage and the disadvantage for the Arduino.

No operating system means that less memory is needed for the micro-controller but it also means that the device can be turned on and off and it will work instantly. The micro-controller can be programmed to run with interrupts and the developer can feel confident that there will be true real time processing of events which cannot always be guaranteed with an operating system. The counter argument is that the Arduino can only be programmed for specific tasks with no flexibility for doing other things.

Raspberry Pi

The Raspberry Pi also came about due to the efforts of Eben Upton, Rob Mullins, Jack Lang and Alan Mycroft from the University of Cambridge's Computer Laboratory. About the same time that the Arduino was getting started efforts were underway to create a tiny computer. Much like the Arduino, the goal of the Raspberry Pi was to create an affordable fun computer for education. Specifically to allow people access to a computer that can be used for experimentation much in the same way the Acorn BBC Micro helped inspire a generation of computer science students.

The expected total market for such a computer at the time was believed to be between ten and twenty thousand units. However, the demand was so great that when it was first available for purchase the planned number of Raspberry Pi's were sold out within minutes and the reseller's, Premier Farnell and RS Components, sites were flooded with so many users that they were temporarily unavailable. When the sites were once again available it was then possible to order a Raspberry Pi but only one per customer.

Although initially a victim of it's own success, the Raspberry Pi Foundation has been able to increase production of the devices up to four thousand per day. This was only possible due to the extraordinary demand that the Foundation actually opened up a second production site in the UK.

The Raspberry Pi specifications are both quite modest when compared to current desktops yet they contain the same horsepower as seen in smart phones from only a few years ago.

Raspberry Pi - Model B

Chip	Broadcom BCM2835 SoC full HD multimedia processor
CPU	700 MHz Low Power ARM1176JZ-F Applications Processor, 32 bit
GPU	Dual Core VideoCore IV® Multimedia Co-Processor
Ram	512 MB
Networking	10/100 Ethernet
USB 2.0	2.0 USB connector
Video output	HDMI
	composite RCA
Audio output	3.5mm jack
storage	SD Card
OS	Linux
Dimensions	8.6cm x 5.4cm x 1.5cm

Table 2: *Original Raspberry Pi specs*

Since the introduction of the Raspberry Pi, quite a few different models have been produced. This also included some reduced specification hardware as well as a few smaller footprint devices such as the Raspberry Pi Zero and Raspberry Pi Zero WiFi.

Raspberry Pi – Model B

Chip	Broadcom BCM2711
CPU	Quad core Cortex-A72 (ARM v8) 64-bit SoC @ 1.5GHz
GPU	Broadcom VideoCore IV
Ram	4 GB
Networking	100/1000 Ethernet
	2.4 GHz and 5.0 GHz IEEE 802.11ac wireless
	Bluetooth 5.0, BLE
USB 2.0	2 x 2.0 USB connector
	2 x 3.0 USB connector
Video output	2 x micro - HDMI
	composite RCA
Audio output	3.5mm jack
storage	microSD Card
Other	H.265 (4kp60 decode)
	H264 (1080p60 decode, 1080p30 encode)
	OpenGL ES 3.0
	Camera Serial Interface (CSI), Display Serial Interface (DSI)
OS	Linux
Dimensions	8.6cm x 5.4cm x 1.5cm

Table 3: *Raspberry Pi 4 specs*

All of the Raspberry Pi's are suitable for using in projects or even embedding.

Operating system

The Raspberry Pi Foundation has several different operating system solutions to choose from for running on the pi.

- Noob
- Raspbian (Debian derivative)
- Pidora
- RaspBMC
- openELEC
- Arch Linux
- RISC OS

These operating systems will perform different duties ranging from a media center for your home up to a complete Linux system for the power user. As I write this all of the available operating systems that are available for the Raspberry Pi are backwardly compatible for the older models of the Raspberry Pi.

The Raspberry Pi is in exactly the opposite position of the Arduino. It is very flexible and can be used for general purpose computing, serving web pages, performing proxy server duties or to support any other computing task else you can think of.

Projects

The Raspberry Pi is also quite popular as a component for DIY projects.

Networking a printer
https://pimylifeup.com/raspberry-pi-print-server/

Home made router
https://hackaday.io/project/4223-raspberry-tor-router

Home surveillance system
http://lifehacker.com/turn-a-raspberry-pi-into-a-cheap-home-surveillance-syst-1474943080

This is possible because the Raspberry Pi is a full featured computer that is available for a very affordable price.

What is IoT
The internet of things is the network of devices, vehicles, home appliances or any other device with embedded software or sensors and is connected to or reachable over the internet. For a long time the trend in computers was that they were both getting smaller and more powerful but the price tag never quite reaching the "right" price for most inventors. Over the past few years the price performance level has finally been surpassed for both manufactured and for casual home projects. The revolution is only possible due to the exponential advances in computing technology but also because of the decreases in the cost and size of sensors and actuators.

The increased intelligence, reduced size and reduced cost have allowed technology to infiltrate all sorts of home appliances. The enhancements started with digital displays but has since added significant technology to the once boring devices such as televisions and refrigerators.

Until not too recently only the large manufacturers have been able to leverage this new technology. The open source community has had the tools such as compilers and integrated development environments as well as the platforms such as Linux and BSD Linux but a desktop computer with these great tools integrates poorly with small projects.

So it was a real boon to the DIY community when the Raspberry Pi and Arduino were released. Between the two devices they can provide the flexibility and the CPU power to support virtually any programming or other embedded task without completely blowing your development budget.

One of the goals of this book is to use or extend a Raspberry Pi or Arduino as well as other tools, devices or sensors to allow you to make your own IoT device or to add intelligence to a common device.

Making your first million
People do not necessarily become do-it-yourselfers because they want to start a company in their garage. The actual reasons are probably more varied and are perhaps as unique and personal as the armies of people who participate in these tasks.

Show off to friends
Impress boss
Desire to produce something physical
Develop skills to change jobs

Create the next Microsoft
Work on different tasks than those from work
Common task with children or spouse
Create a small side job
Take up an interest after retiring
Become an influencer
…

These tools and opportunities make it not only possible but likely that the do-it-yourselfers who produce the most amazing items might not yet shave nor be old enough to drive a car.

It can be difficult to hide laptops, power tools or soldering work stations. My boys have shown some interest and then wanted to participate in these same activities.

Made with free app "Logo Maker" from Worryfree App Studios" by 12 year old

Created using Blender 2.79 by a 14 year old

Created in Microsoft Paint by artistically stunted software developer.

1 THEORY AND COMPONENTS

For software development, a good specification can be worth its weight in gold. When it comes down to it, all the functionality for a piece of software needs to be clearly described but it is indeed rare that everything is described completely and in unambiguous terms.

In electronics this is not the case. Circuit diagrams show the device or component in a way that is unambiguous when viewed and discussed by different engineers. Each of the components have their own symbols as well as their own minimum and maximum ratings and specified tolerances. These engineering specifications are not a simple a book of pictures; These circuit diagrams that are part of the specification that clarify all the parts interactions they are not used for padding out the documentation.

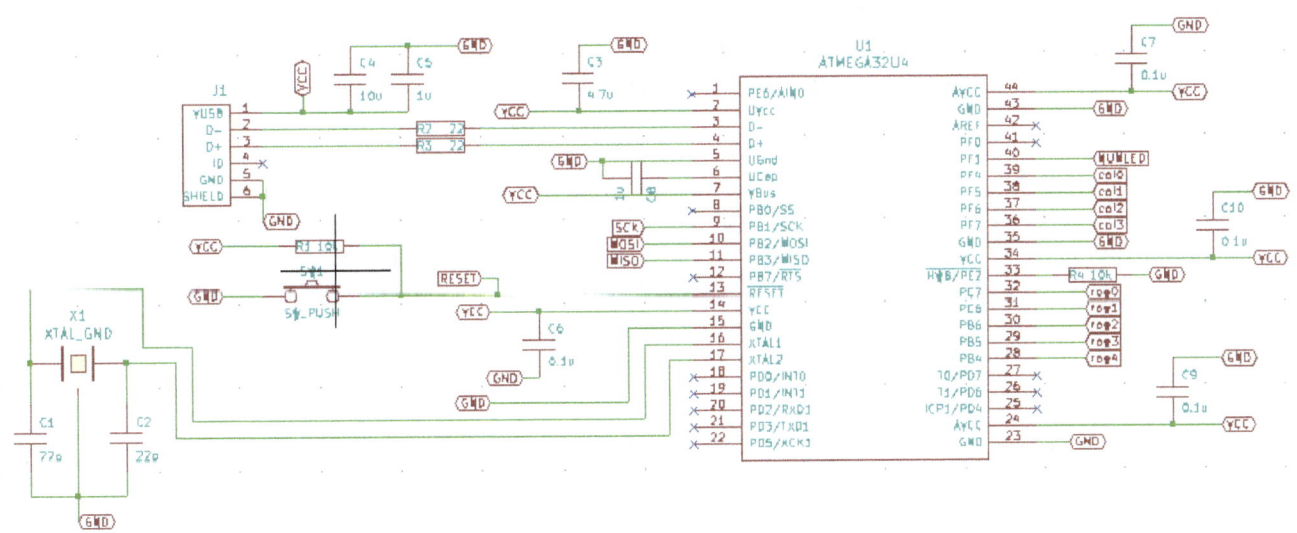

Illustration 1: Sample schematic

To further bolster the clarity of these electrical circuits, is the datasheet. Each manufacturer provides a datasheet which describes the IC or component in truly breathtaking detail. These datasheets describe registers, addresses, timing specifications, interrupts, defined and undefined values or commands, power, graphs but also describe the physical packaging[6] of the component.

Most people take it for granted that when we plug a new lamp into a socket with a light bulb that it will simply work. The reason that it works is because the lamp is expecting a light bulb that conforms to a specific range of power usage, and in turn the lamp has been designed with certain power requirements.

The same is true for electronics as well. Each light emitting diode, LED, resistor capacitor or battery just to name a few have minimums and maximums that they adhere to. When either of these two levels is breached, the parts fail to work, work erratically, have a shorter life or simply stop working.

Roughly said, one of the main reasons for electronic components failing is due to heat. LED's don't normally burn out in the same way that a light bulb does. If too much current passes through the LED then the excess will be shed as heat and if this heat cannot be shed fast enough, then the component will fail due to overheating.

To make sure we don't burn our equipment, we need to ensure that we don't pass too much current through our components. First we should have a greater explanation of what electricity is as well as a few definitions.

Voltage
Voltage is kind of electrical force or pressure from the power supply that is pushing the electrons (current) through the circuit. The units of measurement for voltage is Volts and the symbol V is used in the equations that describe the relationship of the electric field.

When referring to voltage there is often the mention of potential or difference in potential. It is this difference that causes the electrons to flow from high potential to low potential. When electricity was first discovered there was no knowledge of protons and electrons. It was believed at that time that electricity flowed through a wire from positive pole to the negative.

It was centuries after the first discovery of electricity came the discovery of subatomic particles that created a small problem with this explanation. The electron flow did not exactly fit how the diagrams of the time were showing the current flow. It is true that electrons do flow from high density to lower density but the circuit diagrams show the current flowing from the power source (positive) to the ground (negative). The reality is that the electrons flow from the higher concentration of electrons (negative) to the lower concentration (positive).

Although this is the case, the physicists and electricians still show the current flow from positive to negative despite the fact this is not what is happening at a subatomic level.

The voltage can be viewed in a more concrete way with a simple 1.5 volt battery. The difference of potential for this 1.5 volts between the positive and the negative terminals. If we put our multimeter on these to points and measure we should see 1.5 volts on the readout.

Current
Current is the flow of electrons through a wire. It is the movement of these electrons that is referred to as current. The unit of measure for this current is Amperes or Amps. This

---

6   What chip formats, physical size, space between pins, what angles are the legs, or size of the pads.

measurement is just the number of electrons that pass-through a cross section of the wire in a second. It then stands to reason when the current is increased then more electrons are passing through a wire and when the current is decreased fewer electrons pass through the wire.

Current is measure of the number of electrons flowing through the wire, however, what actually makes those electrons move? Electrons move based on differences in potential. Electrons will exist in a piece of wire connected to a LED or light bulb. However, the electrons in the wire do not move in this situation.

How electrons flow is analogous to water. Water does not flow when it is in a pond, however water does flow from a reservoir over a dam to a pond below. The potential in this example is the potential of the water to move without extra energy being applied to the process.

This should be understandable as we know that the water will flow over the dam or waterfall without any assistance (from high to low potential) but cannot flow from the pond up to the reservoir (low to high potential) without putting in addition energy in the form of a water pump.

The same is true for the electrons. The reservoir of electrons would be some type of power source and they would flow through the wire and would end up back at the power source with a lessor potential as seen in Illustration 2. It is the power source that provides the higher potential so the electrons can flow through the circuit.

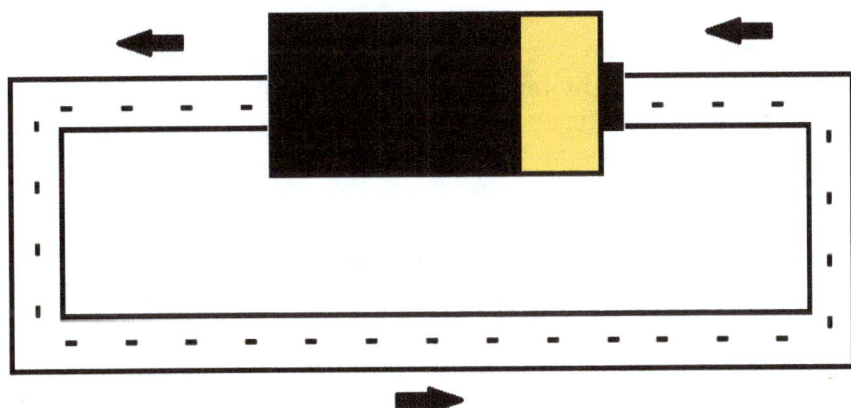

Illustration 2: Electron flow in a circuit

Direct current
Direct current or DC was first promoted by the famous inventor Thomas Edison in the 1880's. Depending on your viewpoint of history perhaps a better clarification would be that the Italian physicist Alessando Volta discovered current in the 1800 and Thomas Edison developed a product which would be used in conjunction with his light bulbs so he could sell lighting systems to businesses.

Direct current is just a steady flow of electrons at an unchanging rate through the wire. When we see electrons constant movement over time graphed we get a straight line such as in Illustration 3.

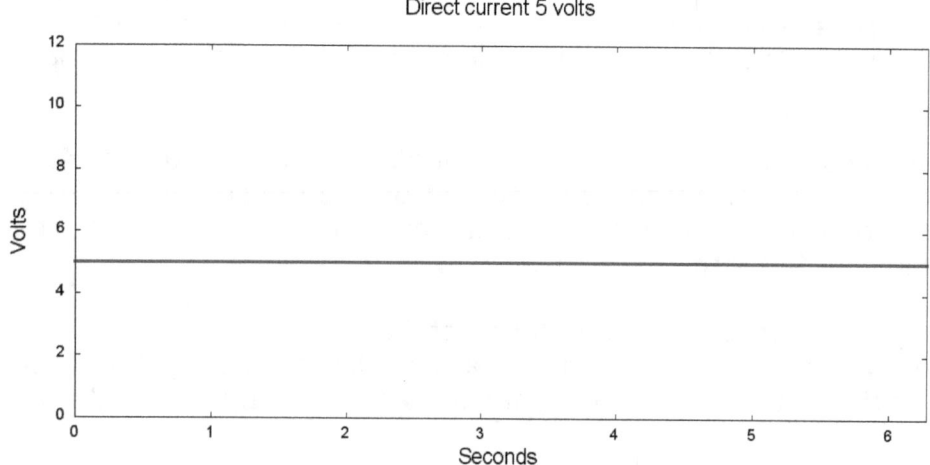

Illustration 3: Graph of direct voltage over time

Certain scenarios have much greater amount of direct current being used, such as an electric automobile while others have much smaller values such as a smoke alarm. In both cases the graph would look identical to this one with only the number of volts being different.

Up until this point we have discussed electricity in pretty static terms. It sounds to be rather fixed at a given voltage and current. For most of the consumer devices we use this is indeed correct. When electricity is delivered at a fixed voltage and current then it is direct current.

When we look at the power charger for our cell phone we may see that it is 5v at 2Amps. The circuitry in the phone is expecting a constant power which is then used to charge the battery. The phone is expecting to receive 5volts and can deliver up to 2 amperes. Some, but not all, devices can utilize smaller amounts of direct current with the expected side effect that less current means slower or dimmer functions from that device.

At the present time, most all home appliances have electronics within the device itself and electronic devices need direct current. Oddly enough homes are not wired with direct current but actually with another type of current called alternating current or AC.

Alternating current

When Thomas Edison was working on his direct current solution there was a different opinion which was held by Nikola Tesla. The solution sold by Edison was to sell power distribution systems for a business. One of the shortcomings for direct current at that time was it could not easily be converted to high voltages. This meant that each business or neighborhood would need to have a power plant nearby to deliver the direct current.

High voltage alternating current made it possible to transmit the power with a lower current and thus reduce the amount lost to resistance. This also enabled the ability to transmit power over longer distances which would then reduce the number of power plants that are necessary.

In direct current solution the current only flows in one direction while alternating current, Illustration 4, changes direction periodically.

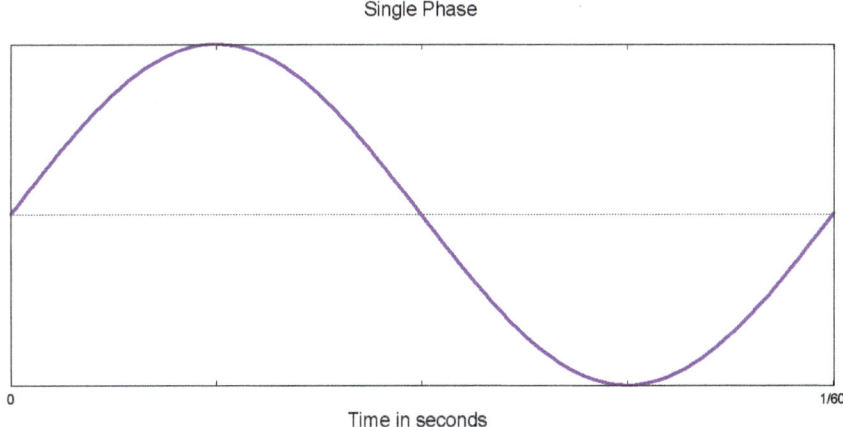

Illustration 4: Single phase alternating current over time

The waveform of the alternating current is a sin wave. The changes is direction occurs 60 times a second in the USA. It is possible to convert between alternating current and direct current but it is only done when truly necessary due to the loses during the process.

Because of the low loss due to resistance, alternating current is used for transmitting the electricity long distances. The alternating current cannot be directly used by a lot of appliances but is instead converted to DC just prior to use. This conversion from AC to DC usually occurs in the power adapter in for small or battery operated devices or in a separate step built into larger devices such as televisions.

It is also possible to create your own AC to DC converter with simple components but may be easier to purchase a ready made power supply. For those interested in the actual theory of how to convert from AC to DC is can read more about it in chapter 4 under the section "Rectifier".

Single and triple phase
The previous section perhaps should have been labeled single phase alternating current. Alternating current has a sin wave form which is very distinctive when compared to direct current. AC changes direction 60 times a second in the USA and other countries but not all countries.

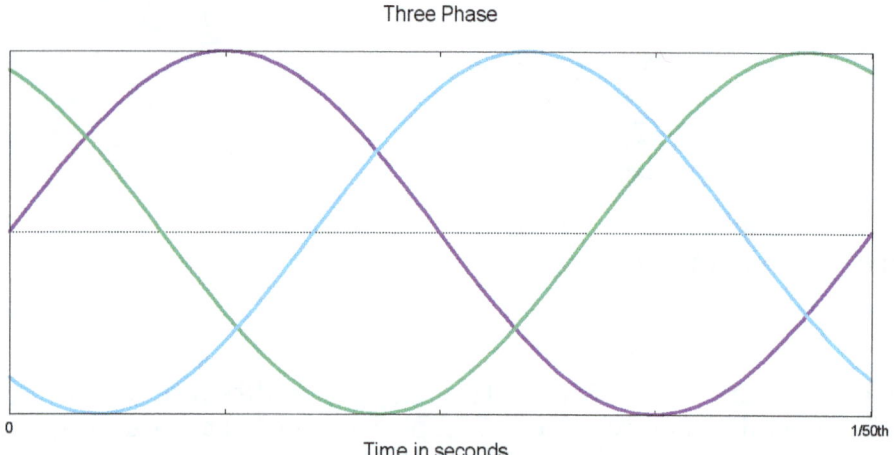

Illustration 5: Three phase alternating current

In Europe the frequency is only 50 times per second. Why exactly are there differing standards in these locations is lost in the annals of time but it is certainly due to the fractured beginnings of electricity with various competing standards.

Yet the frequency of the current in Europe versus the United States is not the only difference between the two. In Europe the power is transmitted in triple phase instead of single. Triple phase, much like the name says, has three wave cycles that overlap. Each of these three cycles are separated by 120 degrees.

Three phase is especially popular in power hungry industries but it is also very flexible. Wiring up a building for single phase has a certain cost but it is cheaper to wire up the same building to support three phase to support the same load. Because three phase is separated by 120 degrees each phase will reach a peak at 0.006 seconds instead of the 0.02 seconds for a frequency of 50hz. This also has the effect of smoother running of motors when using three phase electricity.

Power
Electricity is everything that has been described so far but one of the most useful things about electricity is the ability to store it for future use. This storage is the common battery. However, for energy storage there is no one size fits all and because of that batteries are created in common sizes. What exactly does this mean relative to all the other theory we have covered about electricity?

Electronic circuits need a power source of a certain voltage and current. This combination is often voltage and current when used together is referred to as power. The unit of measurement for power is watts.

$$Power = Volts * Amp \qquad Measured\ in\ Watts(w)$$

This power rating, although interesting, is not the measurement for batteries. The reason for this is there is no indication how long this particular battery would last. This power formula is almost correct. The amendment to the formula is simply adding the number of hours to the equation.

> Power = Volts * Amp * Hours   Measured in Watt-hours(Wh)

Sometimes battery power is not expressed in Watt-hours but instead is expressed in Amp-hours(Ah) or milliampere-hours(mAh).

Energy
Knowing the amount of power that is stored in a battery makes it possible to know how long our electronic device can be powered by a given battery. The formula for determining how long a circuit can run given a suitable powered battery is a simple division.

> runtime = battery power / device usage

Example
Circuit draws   10 mA
9V Battery for   500 mAh

Using the formula we simply divide the 500mAh by 10mAh to determine our circuit could run for about 50 hours.

This analysis is almost 100% correct. Not all batteries are equal and this formula doesn't always hold true. Up until this point has been the assumption that each battery can provide as much power up until its power rating but this isn't the case.

Example 1
Circuit draws   1 mA
1Ah coin cell   1 Ah
Could last   1000 hours
Calculation correct?   Yes.

Example 2
Circuit draws   100 mA
1Ah coin cell   1 Ah
Could last   10 hours
Calculation correct?   No.

These two examples imply our device will run for either 1000 hours or 10 hours. The topic that we haven't really covered is exhaustion.

Batteries can provide power up to the total amount of Amp-hour that the battery is rated for but it cannot all be taken at one time. An example of this is to use a 1Ah coin cell to power a 500mAh circuit. The math says that this battery would last about two hours but the battery doesn't really have the capability to provide so much power for such a long time without becoming exhausted.

The reality is that the battery would last for a shorter time, perhaps even a much shorter time than the two hours calculated by the formula.

The capabilities is determined by the type of battery you are using.

Battery Type	Power density	Pro	Con
Lead acid	7 Wh/kg	Cheap, easy to recharge	Heavy, large
NI-CAD	60 Wh/kg	Inexpensive, easy to recharge	Contains toxic metal
Alkaline	100 Wh/kg	Safe, long shelf life	Not rechargeable
NI-MH	100 Wh/kg	High power	Self discharge fast
Lithium Ion	126 Wh/kg	Ultra light, high power	Can explode
Lithium Polymer	185 Wh/kg	Ultra light, high power	Can explode
Lithium Coin cells	270 Wh/kg	High density, small	Low current draw

Table 4: *Various batteries and their attributes*

It is safe to say that lithium ion coin cell batteries cannot support too high of a draw for too long else they will diminish faster than calculated. There is no one size fits all values for the capability of the batteries. This information must come from the technical specifications for each battery.

It is possible to use larger or more batteries to power a device this is not always very convenient. For a given device a battery could be designed to provide 10 hours with a certain current however if we could somehow reduce the current by half, we would be effectively doubling the run time with that same battery. Using a resistor can make this a reality for certain devices especially those that are primarily LED driven. If we limit the current to our LED then our LED based device will light up for a longer time, although it will be not as bright as before.

Resistance

At the playground sliding down a pole can be faster or slower depending on your clothing and how tightly you are holding onto the pole. The holding on tight is the main source of resistance which will slow your decent.

In electric circuits resistance is not much different than other real world situations. Lots of devices need electricity but in the same way people don't drink directly out of a fire-hose these devices also need a lessor amount of current.

The current is controlled by adding components that prevent (resist) the flow of electrons to a manageable amount for that circuit. These components are called resistors whose jobs is simply the prevention of electrons flow. Resistance is measured in Ohms and uses the symbol Ω.

Ohms Law

The voltage (V) across the resistor is proportional to the current (I) where the constant proportionality is the resistance (R).

$$V = I \times R$$

However, it is fairly typical to restate the equation solving for the current

$$I = \frac{V}{R}$$

This law basically says increasing the resistance while keeping the voltage the same will allow less current through that part of the circuit. This is as important for electronics as the components are rated in how much voltage and amperage (current) they are rated for.

Read and re-read this, as this formula will come back again and again when calculating for various components.

Practical Example

Up until this point has been all theory. The question is how exactly do you use these formulas to select resistors.

$$V = I \times R$$

To make this concrete we need to know both the input power but also more information about the circuit. For this example we will use a circuit that is essentially power and a LED.

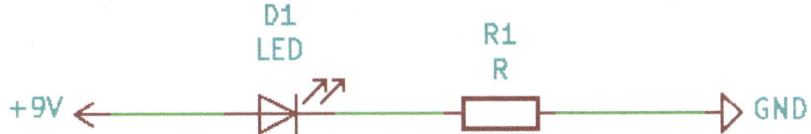

Illustration 6: Circuit that is only power to drive LED

Some constants

LED
Color   Pink
Vf   2.2volts
Current   20 mA
Brightness 400mcd

Power
V   9 volts

What resistor do we need? We start by doing a bit of formula manipulation to solve for R.

$$R = \frac{V - V_f}{I}$$

Then we plug in the values from our parts.

$$R = \frac{9v - 2.2v}{20\,mA}$$

Calculate the voltage after the forward voltage from the LED is subtracted. In addition, convert the number of milli-Amperes to Amperes.

$$R = \frac{6.8v}{0.02\,A}$$

Finally, solve the remaining value to find out the size that the resistor needs to be.

R= 340 Ω

For this particular circuit we would need a 340 ohm. This would be the recommended size for this circuit but it is possible to have either a larger or smaller resistor. If a smaller resistor is used then the circuit will still work. The downside of a smaller resistor is the LED may stop working after a while because the LED cannot shed the excess heat.

A larger resistor can be used and will have the opposite effect. The LED will receive less power and will be somewhat dimmer than when it receives the maximum power.

Components

Wire

Wires are measured in diameter, and of course there are various different standards for measuring them. The standard in the United States is American Wire Gauge (AWG) which is a measurement of the diameter of a raw wire without any insulation. Because of the international world we live in table 5 shows the AWG values in both imperial and metric sizes.

Perhaps contrary to what might be expected, the higher the gauge number actually represent the smaller the diameter for the wire. The thicker wires have less resistance over the length of the wire and thus are better for longer distances.

Typical household power wiring is using 12 or 14 gauge wires while telephone wiring may be 22, 24 or 26 gauge.

AWG	Diameter[7] (inches)	Diameter (mm)	Cross sectional[8] (mm)	Amps
30	0.010	0.2540	0.051	0.134
28	0.013	0.3302	0.086	0.213
26	0.016	0.4064	0.130	0.338
24	0.020	0.5080	0.203	0.538
22	0.025	0.6350	0.317	0.856
20	0.0319	0.8102	0.516	1.36
18	0.0403	1.0236	0.823	2.2
16	0.0508	1.2903	1.308	3.4
14	0.0640	1.6256	2.075	5.5
12	0.0808	2.0523	3.308	8.7
10	0.1018	2.5857	5.251	13
8	0.1284	3.2613	8.354	22
6	0.1620	4.1148	13.298	35
5	0.1819	4.6202	16.766	44
4	0.2043	5.1892	21.149	55
3	0.2294	5.8267	26.665	70
2	0.2576	6.5430	33.624	88
1	0.2893	7.3482	42.409	111
0	0.3248	8.2499	53.455	140

[7] 1 inch = 25.4 mm

[8] mm cross section. Area = $\pi \times r^2$

Table 5: *Wire dimensions in imperial and metric sizes*[9]

Note: The current carried is based on the cross sectional value. A single square mm can carry between 4 and 6 amps.

** It is better to error on the side of caution and that perhaps using a gauge of wire that is larger than your calculated amps would be the prudent thing to do.**

Despite what all the information thus far, none of this really describes one additional fact about wires. They can be either single strand or multi-strand but why should one be used over the other. Actually, is there a difference between the two? Either single or multi-strand wire of the same gauge can be used but there are pros and cons.

Advantages	Disadvantages
The wire is usually smaller at the same gauge.	Larger gauge wire can be harder to use as it is more difficult to bend.
	Flexing and bending too often will fatigue the wire and can cause it to break.

Table 6: *Solid wire*

Advantages	Disadvantages
They are less likely to break from a small nick.	Larger gauge wire can be harder to use as it is more difficult to bend.
They are less likely to suffer from metal fatigue.	Flexing and bending too often will fatigue the wire and can cause it to break.
They are easier to move or route as they bend more easily.	

Table 7: *Stranded wire*

It should not make any difference which type of wire you select for small electronic projects as long it is suitable for the current it will be carrying. However, that having been said, stranded wire tends to be a better choice if it is possible that the project will be on the move – e.g face some vibration.

It is the constant vibration that may fatigue the wire causing it to break over time. This isn't so much a problem with building wiring as the building has a tendency to not move much nor often.

Breadboard
This is perhaps one of the most convenient pieces of hardware that can be used for your projects. It is easy to connect LED's, chips, switches, wires, resistors or just about any other button or part. What makes this so great is that it can be done without any tools

---

[9] Online information about http://www.ken-gilbert.com/techstuff/AWG_WIRE_TABLE.html

whatsoever.

The board is designed to allow chips to be plugged into the center giving access to the legs on either side but at the same time still keeping both sets of legs electrically separate from each other. The spacing of the holes is 0.1 inches (2.54mm) which is also the exact spacing required for chips in the dual in-line package (DIP) format.

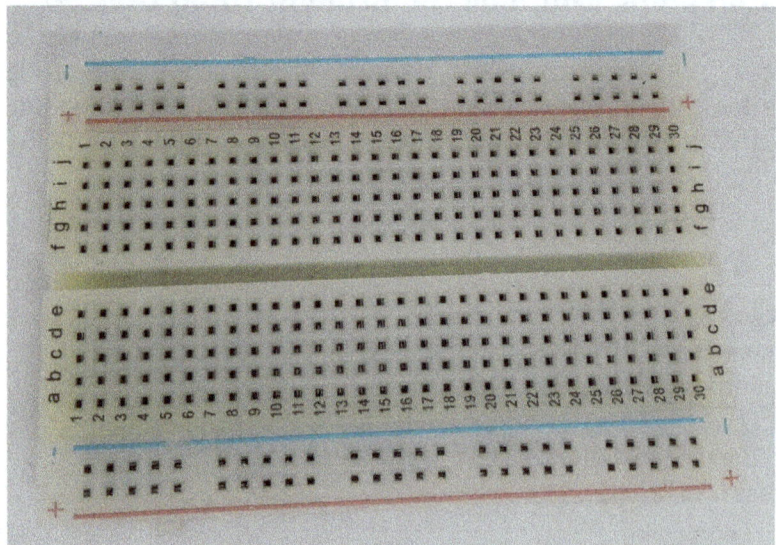
Illustration 7: Breadboard

Breadboards are also very convenient for prototyping projects prior to soldering it together. There are two different general styles of electronic parts. They are through hole and surface mounted device parts.

Through hole parts are parts that have leads that stick out of one end or both depending on the part. These legs can be bent and then pushed through a hole in the circuit board. There are quite a few different types of parts that may be through hole parts such as a resistor, capacitor, diode, integrated circuit (IC), relay, LED, or transistor. This is by no means a definitive list of components and as luck has it each of these components have multiple different packages as well as various sub-types of components.

The advantage of through hole components is that the integrated circuits in DIP format can be plugged directly into the board as this format also uses 0.1 inch spacing for the pins. All other through hole components have legs that can be slightly pulled apart when inserting them or can be bent to easily realize your board layout - e.g a resistor or capacitor can be made do connect two different rows that are not adjacent.

Prototyping board
A device whether Raspberry Pi, graphics card for your pc, or homemade project may use anywhere from a single microprocessor or memory chip upwards of dozens of resistors, capacitors, LEDs, or other integrated circuits.

Despite how important the components are there is one item which shouldn't be overlooked is the circuit board that these components are mounted on. It is easy to take a casual glance and think that they are all the same but there are two common types of prototype board that are actually very different.

Printed circuit board history
A long time ago computer programs used to be fed into the computer with punched cards. This is ancient history which usually gets perhaps one or two sentences in courses about the history of computing. It isn't only software that sprung from creaky old technology but also hardware.

Hardware started out with vacuum tubes and lots of wires with the construction being point to point construction mounted on a board made of varying materials. Eventually the circuit boards went from a single sided constructs mounted to wood but eventually morphed into boards made out of resins and fiberglass. The materials for the circuit boards is a fiberglass board with a copper platting on one side.

The actual process is more complicated but essentially desired wiring would be printed on the copper side of the board in an acid resistant ink. The rest of the copper without the ink is then removed with an acid solution. Then required holes for any of the through hole components were then drilled.

Over the next few decades the process continued in this manner with the technology improving the process of the transfer of the printed design. During this time other factors also came into play such as board with multiple layers and surface mounted parts. This allowed both smaller and more complex circuit boards to be created.

The technology didn't only improve the process of transferring the printed design during the creation of the printed circuit board. The improvements also increased the precision of the design allowing smaller traces with a width of 4.5 - 6 mil which also allowed higher density boards.

Somewhere during the development of the common printed circuit board it became obvious to hobbyists that this method of board creation could be performed by anyone. Even today it is possible to find information on the internet how to create your own printed circuit board. There is a limited number of steps and none of them are terribly complicated.

> Etching chemicals
> Plastic trays for chemicals
> Blank copper boards
> Laser printer for creating design
> Safety gear (goggles, gloves, filtered mask)
> Dremel or drill press
> Jig saw or hack saw

None of these steps are tricky but the chemicals must be used in a well ventilated location due to their toxic nature. Not only do you need to be careful of the fumes but you also need to find a manner for disposing of the used chemicals which are toxic to fish and other water based organisms.

Other than this particular drawback, it does sound like etching your own boards might be fun for simple projects. Due to the other options available, etching your own circuit boards will not be discussed any further.

Single sided prototype PCB
The most common prototype board that I have seen is tin with copper on one side. This

copy may be a small circular pad around each hole, a longer pad that goes around three holes, or may be copper strips from one end of the board to the other connecting all holes together for that row.

Depending on your source, this common prototype board is relatively inexpensive and readily available on the store shelf. There is nothing wrong with using this type of board for projects but there are a few facts that should be known up front.

Illustration 8: Single sided copper prototype board

The copper will oxidize as it is exposed to the air, so it is important that you use your flux to remove the oxidation when soldering. The copper is attached only on the surface of the board so while soldering you must be careful, too much heat and pressure from the soldering iron may shift or remove this copper plating. Once the soldering has been completed, some care must still be taken as the components are soldered to this copper plate, it is possible that with poor handling that the component and copper plate can come loose.

Double sided prototype PCB
The second type of prototype board is typically green with a silver looking pad around each hole. In this sense, it is very boring but there is one very great difference. This pad is not simply attached to the surface layer of the board but connected through the via[10] with metal tube to a pad on the other side of the board.. Thus the metal starts on one side, is on the sides of the hole going through the board and connects to the other side. Because of this type of construction, this piece of metal is quite stable and unmovable.

---

10 A via is simply the path connecting the two different layers of the PCB board. This can also be connecting through from one side to the other.

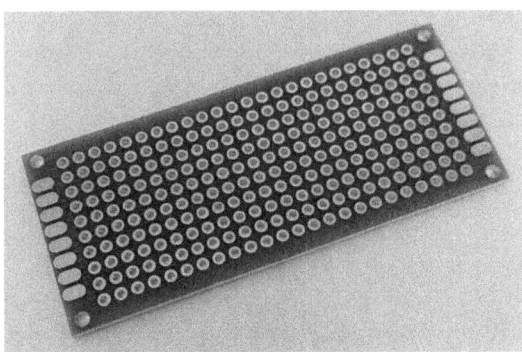
Illustration 9: Prototype board with plated through holes

There is one other benefit, which is it is really easy to solder a solid connection. When soldering a wire or component to the board you will place the iron tip on the pad touching the wire and then carefully touch this junction from the other side with your solder. The solder will melt and flow into the hole forming a very solid connection.

Light Emitting Diodes
A LED is a semiconductor light source. It can be used for generating lights in the home or indicating whether an electronic device is on. It only lights up if power is applied to the anode and the ground is applied to the cathode.

Unlike a light bulb where the maximum watts are displayed on the bulb itself, the LED usually has none of this. When purchasing LEDs there will be specification sheets which will describe the limits, but unlike light bulbs, many LEDs look very similar but do have very different power requirements. The are rated by both the maximum voltage and the maximum amps that the LED can withstand. This is usually between 2.2 and 3.6 volts and 20 to 30 milliamperes.

Different LEDs have different specifications regarding how much forward voltage and current used. However, as rule of thumb you could use the following voltages assuming 20 milliamperes.

Red	1.8 V
Green	2.1 V
Blue	3.3 V

These of course are just estimations, as there are also low power LEDs which use as little as two milliamperes current.

Just like light bulbs, light emitting diodes can only accept so much current before they break, part of the current that passes through them is converted to light while part of it is converted to heat. It is the increasing heat from power dissipation will cause the LED to fail when more current is passed through it.

The lifetime of a LED when properly treated with electricity in the proper range and with proper cooling can last tens of thousands of hours.

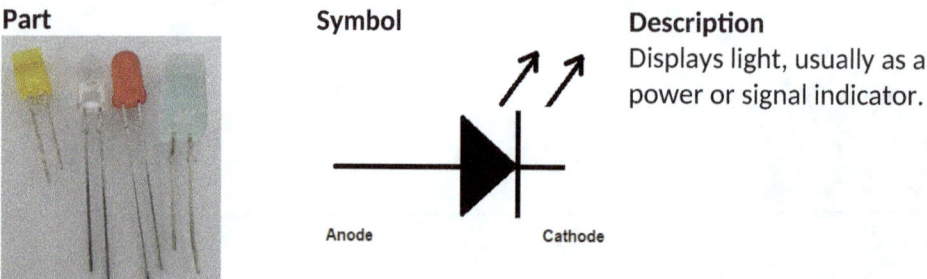

Part	Symbol	Description
	Anode — Cathode	Displays light, usually as a power or signal indicator.

In order to limit the flow of electricity to our LED or other electronic components, we need a material which provides an easily measured way to prevent the electrons flowing through our circuit. An electronic part designed for such a task is the humble resistor.

Resistors

A resistor can be made of virtually any type of material that conducts electricity. The idea is that the resistor should control, typically to reduce, the current by providing resistance. They are usually made from a material which has some conductance such as carbon and then mixed with other components that do not conduct such as clay. Different proportions of these materials will create different resistances.

A wire is inserted from both ends and due to the less than ideal conductance the electron flow is reduced. Theory may be simple, yet the manufacture is undoubtedly more challenging. The good news is that it is not necessary to make our own resistors they can be purchased in varying resistance levels anywhere from less than 1 ohm up to millions of ohms.

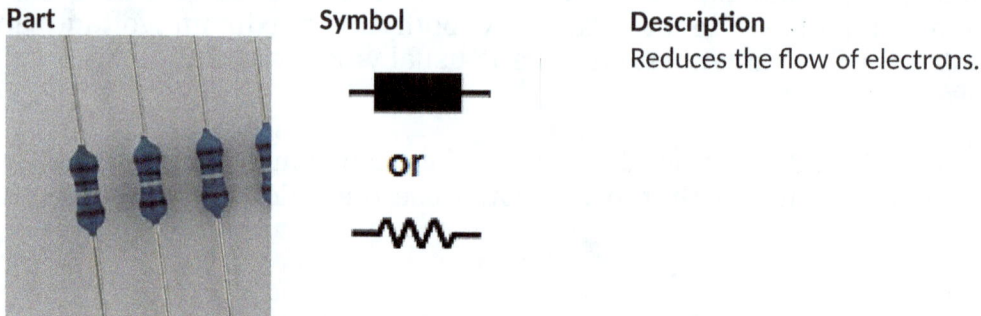

Part	Symbol	Description
	or	Reduces the flow of electrons.

All electronic parts come with technical specifications which would be the right place to determine what it is capable of. With resistors it may not even be necessary to read the spec sheet. Resistors are color coded, so it is possible to determine their resistance by sight. Simply look at the colors of the resistor and using the chart will allow you to determine the resistance of the resistor.

Illustration 10: Through hole resistors color example

Color	1st digit	2nd digit	3rd digit	Multiplier	Tolerance
Black	0	0	0	1	
Brown	1	1	1	10	1%
Red	2	2	2	100	2%
Orange	3	3	3	1000	
Yellow	4	4	4	10k	
Green	5	5	5	100k	0.5%
Blue	6	6	6	1M	0.25%
Violet	7	7	7	110M	0.1%
Gray	8	8	8		
White	9	9	9		
Gold				0.1	5%
Silver				0.01	10%
None					20%

Table *8: Through hole resistor color table*

Having this color coding system simplifies the identification of how much resistance any "unlabeled" resistor may have. There are also a couple of other fairly easy methods for identifying this resistance. The first is to use a multimeter to simply measure the resistance, it is a relatively inexpensive tool that you will probably have around. A more high-tech solution is to get application for either your computer or smart phone.

It is a bit disingenuous to suggest you do not need to read the spec sheet, as resistors are rated as to how much power they can handle. You would not want to push too much current through too small a resistor as that would cause it to get hot and perhaps even burn.

Typical ratings for resistors in electronics devices are 1/8, 1/4 and 1/2 watt. You should simply be aware of this when selecting your components. You can calculate how much power your resistor will need by the following formula depending on which values you have available.

$$P = I^2 \times R \quad \text{or} \quad P = \frac{V^2}{R}$$

While reading data sheets or looking at circuits or project kits there may be a bill of materials listing all the components. It is easy to recognize the resistor when it simply shows the amount of resistance followed by ohms. However, as the resistors get larger the syntax is slightly different depending on the source. It might be either 4.7k or 4k7 can be used to represent a 4700 ohm resistor. For resistors in the million range will use M such as 4MK.

The perfect resistor
If we are using a resistor to ensure we don't pass too much current to our LED we can choose anywhere from a minimal value up to a more conservative value. What would a minimal value be? We would want a resistor that will ensure that the forward voltage does not get exceeded.

Using a higher rated resistor doesn't prevent our LED from lighting although it may prevent it from lighting to it's maximum potential.

We can use Ohm's formula to calculate how large a resistor to use. My blue LED has a forward current of 20mA and a forward power of 3.8 volts.

$$V = I \times R$$

We simply solve for R as we know what our power source is.

$$R = \frac{V}{I}$$

Our voltage is 5V and the LED can take up to 20 mA of current and 3.8V so plugging these values into our formula we come up with a 60 ohm resistor.

$$R = \frac{(5 - 3.8)}{0.02}$$

$$R = 60$$

Well, should we take this value and simply move to the next step? No. From our chart on the previous page we can see that the tolerance can be off by as much as 20% depending on the resistor purchased. So the 60 Ohm resistor may actually behave like that of a 48 or a 72 ohm resistor.

Should we use a 60 ohm resistor? Theoretically yes. However these are the maximum values and using the minimum resistor value will maximize the brightness of our LED. Why does this matter, well, what is more important to protect the ten cent LED, a fifty cent IC or the Raspberry Pi? While we can use smaller resistors in theory the goal at least at the beginning is to also protect the GPIO of the Raspberry Pi. Thus instead of a perfect resistor perhaps a safer resistor might be a 330 Ohm or higher, but smaller resistors can be used.

However, before we leave resistors, there is one more device which behaves similar to a resistor. The device is called a potentiometer and allows the resistance to be changed by turning the knob. potentiometer's are simply resistor with a range, the start at zero and go up. A low end potentiometer would be like 0 - 10k or 0 - 50k.

Part	Symbol	Description
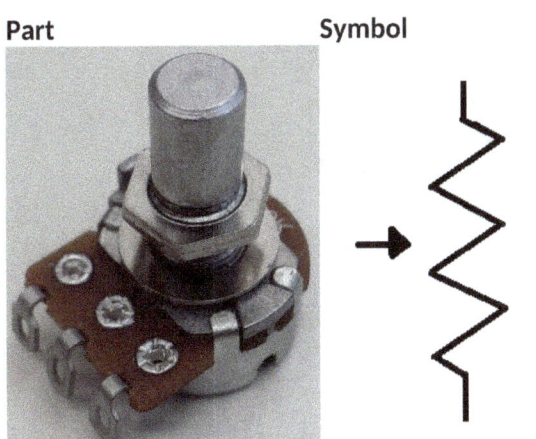		Like a resistor it reduces the flow of electrons, however, it can be turned to any value that is supported by that potentiometer. e.g. 0 – 10k ohms

Part	Symbol	Description
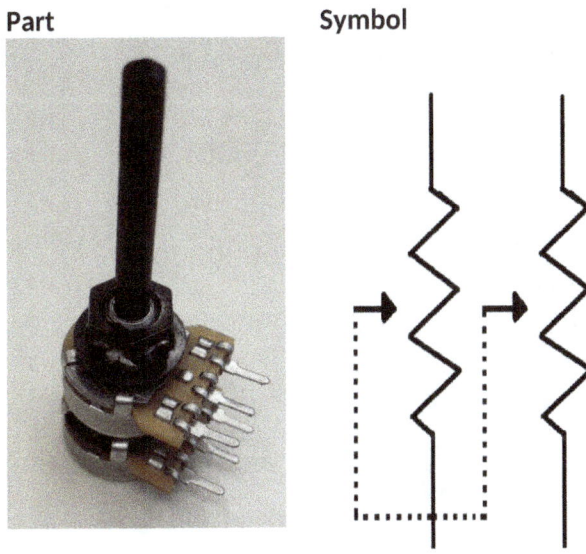		Much like the single potentiometer this dual potentiometer will allow you to select a resistance somewhere in the range supported by the device. One difference is that this will allow you to vary the resistance for two separate inputs using the same value.

There are even potentiometers that take multiple sets inputs and apply the same resistance to both sets. One such example might be a volume control on a stereo headset.

Capacitors

What are capacitors? It is a device with two electrical conductors separated by an insulator that stores charge. When there is a difference of potential between the conductors then an static electrical field develops across the insulator. Positive on charges on one plate and negative on the other.

Part	Symbol	Description
		Polarized capacitor. Stores energy in an electric field, must be properly inserted into circuit otherwise it will fail or worse. Typically this is the case for large capacitors.

Part	Symbol	Description
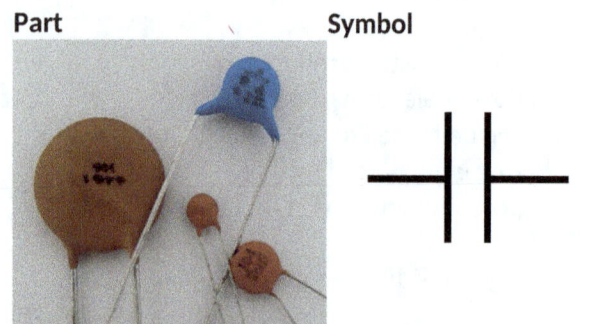		Non-Polarized capacitor. Stores energy in an electric field. Orientation in circuit not important.

Capacitors are measured in Farads (F) after the English physicist Michael Faraday. A farad is the charge in coulombs which a capacitor will accept for the potential across it to change 1 volt. A coulomb is 1 ampere second.

1 millifarad	= 1mF		one thousandth of a farad (0.001)
	= 1000mF		one thousand microfarads
1 microfarad	= 1mF		one millionth of a farad (0.000001)
	= 1000nf		one thousand nanofarads
1 nanofarad	= 1nf		one billionth of a farad (0.000000001)
	= 1000pf		one thousand picofarads
1 picofarad	= 1pf		one trillionth of a farad (0.000000000001)

Note: Capacitors do not charge in a linear manner.

Why are there both polarized and non-polarized capacitors. The why actually has more to do with the physical make up of the capacitors themselves.

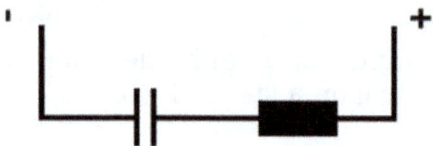

Polarized capacitors are small with a large capacitance. They are used in direct current circuits. Non-polarized capacitors are actually physically tiny and support a fairly small capacitance. They are used in both direct current circuits and alternating circuit circuits as well. Often they are used as a bypass capacitor which is used to filter out noise from the power supply. In general these two classes of capacitor cannot be used as replacements for each other.

The capacitor will charge based on the RC time constant of the circuit. This value, tau, is the product of the resistance in ohms and the capacitance in farads of your components.

$$\tau = R \times C$$

This constant is the amount of time it would take to charge the capacitor by 63% from the initial to the final voltage. The formulas for the charging and discharging and a graphical representation is shown below.

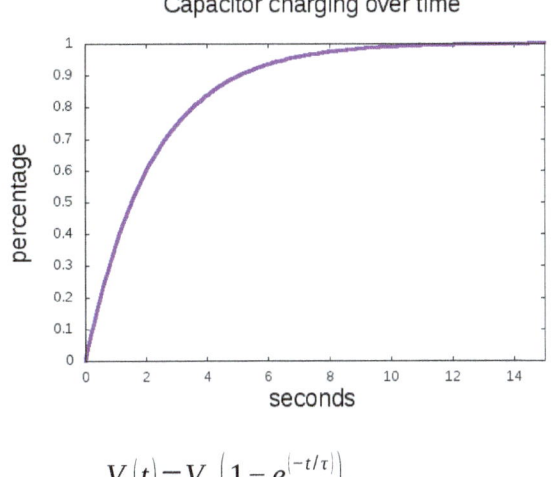
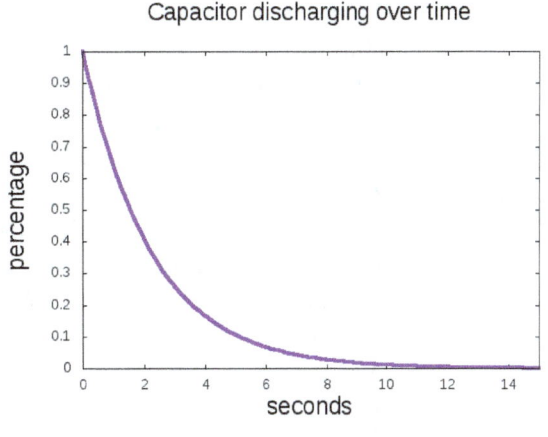

$$V(t) = V_0\left(1 - e^{(-t/\tau)}\right) \qquad V(t) = V_0\left(e^{(-t/\tau)}\right)$$

Unfortunately charging a capacitor, it is like when my math teacher told us that we could leave class right now, but we could only move ½ of the distance between our chair and the door every minute. In practice you could get close to the door but in theory there would always be a small distance left to cover.

The capacitors will only charge 63% in one time period, and then 63% of the remaining amount in the next time period and so on. However, in practice a capacitors will be 99% charged after 5 time periods.

Time	Amount charged	Total charged	Remaining uncharged
0	0	0	1000
1	630	630	370
2	233	863	137
3	86	949	51
4	32	976	24

R = 2200 ohms
C = 1000 mF

t = 2200 x .001
  = 2.2 seconds

In this example, it would take eleven seconds to charge up this capacitor to 99% of its value.

Note: This does not mean you can speed up the charging process by eliminating the resistor. Depending on your power source you could blow up the capacitor.

Types of capacitors
There exist quite a few different types of capacitors. Unless you have special needs, you may not actually encounter more than one or two different types or any of their different package formats.

Type	Description	Range
Electrolytic	Large value capacitors used in both power filtering as well as to decouple one part of the circuit from another part. These capacitors are polarized, meaning they have a positive and negative leg.	1µF and above
Tantalum	Made of tantalum metal. They have a very high capacitance per their size, low current leakage, and long term stability. Like the electrolytic capacitors they are also polarized.	0.1µF to 1000µF
Ceramic	Used in power filtering	1pF to 1nF
Film	Used in tone controlling as well as power filtering.	1nF to 1µF

The different capacitors come in different formats. The electrolytic look like a small can, with the legs coming out of one end. Although less common, there are also electrolytic capacitors that look more like a resistor as one leg will come out of each end.

The ceramic capacitors look like a small disk with the legs coming out of one side. The formats for the other capacitors can vary. They may look somewhat similar to a ceramic capacitor, or they may be rectangular in shape.

Uses for Capacitors

DC Blocking	Capacitors can be used to allow alternating current through yet it will block the direct current from passing through.
Power smoothing	Capacitors can be used to filter out noise or spikes power that is being fed to the circuit.
Audio Filtering	Capacitors can be used for filtering audio signals in much the same way they are used for power smoothing.
Timing	Capacitors when used with other components and integrated circuits can be used to create a timing circuit.
Energy Storage	Capacitors do not have the same storage capability that a battery has, yet they can be used to store reasonably small amounts of energy for release. This energy can be released at once or over time depending on the application.
Converting AC	Capacitors can be used with a rectifier to convert alternating current to direct current.

Note: A final word of caution, not all parts of the body react kindly to the current associated with electricity, starting with fingers but not excluding lungs or heart. Ten milliamperes will produce a painful shock and as little as one hundred milliamperes can easily cause death.

Note: Capacitors store the current that can be so deadly. This is especially true for larger capacitors that come out of some large electronics items such as TVs or

other large consumer electronic devices. Be careful that you do not simply connect the two pins to discharge it as depending on the quality of the capacitor it may get hot or explode.

Final word on capacitors
Electrolytic capacitors are slightly different from other types of capacitors as they contain fluid electrolytes. These capacitors work just fine but if problems do occur they can be due to the capacitor leaking.

Illustration 11: Example of capacitor having leaked onto the circuit board

There are two different visual ways to see problems with electrolytic capacitors. In illustration 11 it is possible to see that this capacitor failed and has leaked its all over the circuit board and covers the resistor as well as one of the leads for several of the diodes. A second indicator of a blown capacitor is that the top will bulge.

Serial and Parallel
Before covering any other components I want to discuss a little bit more about these two concepts and how they sometimes used in circuits. The components can behave somewhat differently depending on how they are assembled. What we mean by that is when the components are placed one after another, e.g serially, or all next to each other in parallel .

Serial
If you are old enough, you may remember Christmas tree lights that when one bulb burned out the entire string would not light. This is an example of a serial circuit.

In a word, well a couple of words. This circuit can be thought of as a single wire, where the electricity flows through each component and if any single component fails, this is equivalent of a small break in the wire, thus no current gets through and nothing will light up. The solution for this string of Christmas tree lights is obviously you need to find the bulb or bulbs that failed – a tedious task.

One of the attributes of a serial circuit is just this example, when one component breaks that prevents the operation for the rest of the circuit. This should not be too surprising as we would probably expect poor if any operation where one or more of the components break.

Yet this is not the only example of a serial circuit in daily life. Another example of this is the common flashlight. Most flashlights have a single bulb but have two or more batteries arranged serially.

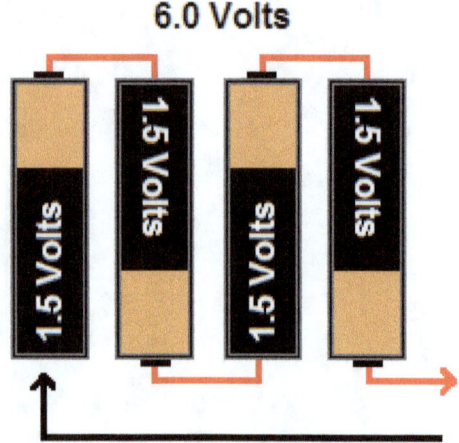

Illustration 12: Batteries in series

It is the serial placement of the batteries that yield a slightly different behavior of our standard 1.5 V AA battery. The voltage is 1.5 volts yet when they are put in circuit serially the voltages add up while keeping the current the same[11].

Putting batteries in a series allows us to basically create any higher voltages by simply adding more batteries to our device. We can create 3, 4.5, 6, or even 7.5 volts batteries by combining a number of 1.5 volt batteries in this manner.

Resistors in a series is actually quite similar to our battery example. The resistance is cumulative, simply adding more resistors to a circuit in series adds to the total resistance.

$$Rtotal = R1 + R2 + R3 + ... + Rn$$

Actually for resistors this is even more flexible as it is possible to add any number of any resistance together, while for batteries it is best to only put batteries in a series that have identical voltages and amperage hours together.

Capacitors behave completely differently and are perhaps a bit non-intuitive when put into a series. The total capacitance of capacitors in a series is equal to the sum of the reciprocals of each capacitor.

11 This is assuming that the batteries are all identical.

$$\frac{1}{Ctotal} = \frac{1}{C}1 + \frac{1}{C}2 + \ldots + \frac{1}{Cn}$$

Parallel

A parallel circuit is a closed circuit in which the current divides into two or more paths before it is recombining to complete the circuit.

This actually is the perfect description of a string of Christmas tree bulbs[12]. Well, actually most Christmas tree lights these days are really powered by LED's. Actually it doesn't really matter if we are talking about lights or LEDs it is the construction that is important. As the definition says, the circuit is dividing into multiple paths, in this case, one for every light. If that component fails, the other paths still complete and cause the rest of the lights to light up.

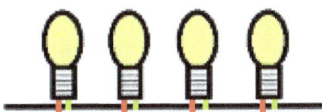 

This is really good news for Christmas tree's everywhere, and it also is an example that indeed the same components when rearranged produce some really different results.

Reexamining the common battery but in a parallel setup we see not the voltage that is increasing but the current. In this case, the current is multiplied by the number of batteries that are connected together.

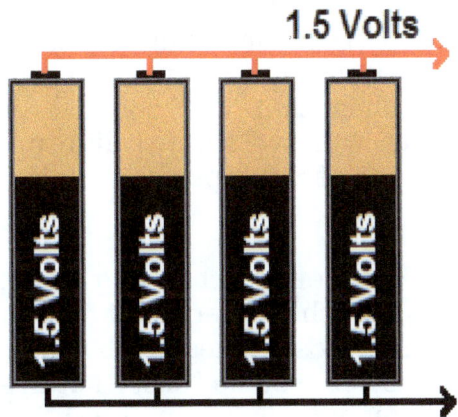

Illustration 13: Batteries in parallel

This will allow the circuit powered by these batteries can either consume a larger current or to last four times as long.

As far as the calculations go, the resistors and capacitors have switched positions. The calculation for resistors that are in a circuit in parallel actually calculated in the exact same manner as we did for capacitors in a serial circuit.

---

12 Actually most sets of Christmas lights now are really powered by LEDS.

$$\frac{1}{Rtotal} = \frac{1}{R}1 + \frac{1}{R}2 + \ldots + \frac{1}{Rn}$$

To get the total resistance of our resistors that are in parallel again we add the reciprocals of each of the resistors. The total resistance of these resistors in parallel will always less than the resistance of the smallest resistor.

It is not possible to make the statement that this type of circuit would never be necessary but it is perhaps less often used compared to simply selecting the proper resistor for your circuit.

However, putting capacitors in parallel is a more interesting situation. The total capacitance of capacitors in parallel is equal to the sum of all of the individual capacitors.

$$Ctotal = C1 + C2 + C3 + \ldots + Cn$$

This allows us to take a number of smaller capacitors and essentially create a larger capacitor. However, the voltage of the this "pseudo" capacitor is equal to the lowest voltage of any of the capacitors in parallel.

Switches

As obvious as it might sound perhaps switches should be covered. Perhaps the most common household switch is a light switch. Flipping the switch will complete the circuit thus turning on the light. This type of switch is called a single pull single throw (SPST).

The "poll" represents how many contacts sets there are, a single pull single throw has one set of contacts while a single pull double throw has two sets. There are many different types of switches available. There are also switches that will switch between two different connections such as single poll double throw (SPDT), switches that will connect or disconnect two different circuits (DPST) or even switches that will switch two different circuits to their other value (DPDT), but this is not a definitive list of the types of switches available.

As previously mentioned, these switches are actually permanently changing flow. However, the thing that distinguishes a switch is that it latches, which is to actually lock to other setting. If the item doesn't latch then these are not necessarily switches but buttons. The two basic types of buttons are those that are "push to make" as well as those that are "push to break". make of which may make or break contract momentarily when pressed.

Part	Symbol	Description
SPST		Single poll single throw

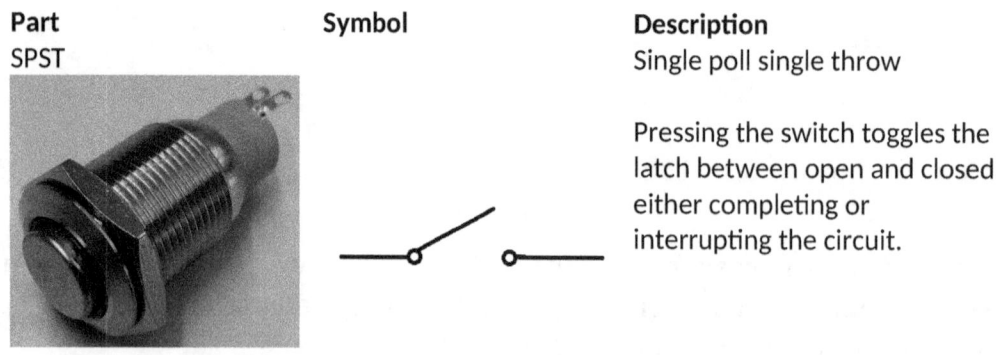

Pressing the switch toggles the latch between open and closed either completing or interrupting the circuit.

Part	Symbol	Description
Button 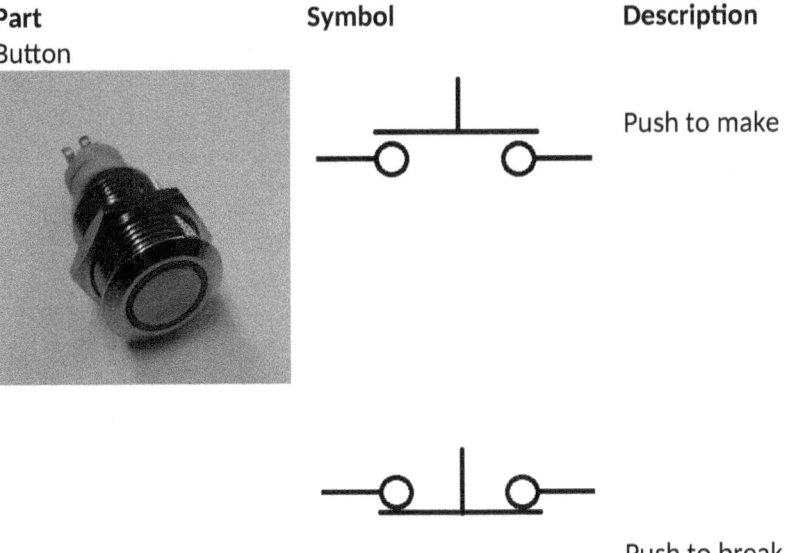		Push to make
		Push to break

Transistors
This is a general term to cover a lot of individual parts the commonality is that all those parts are semiconductor devices.

Bipolar Junction Transistors
These are used for amplification of the voltage as well as switching of the signal. There are indeed technical differences such as which direction your current is flowing however they are not significant for the simple circuits that will be described in this book.

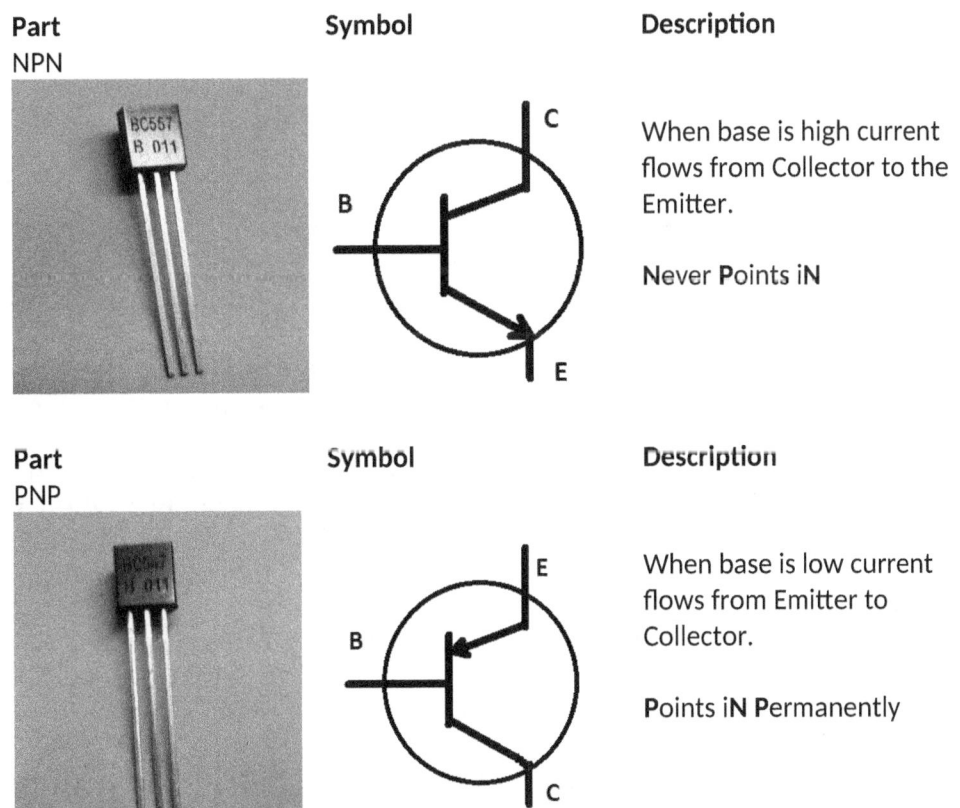

Part	Symbol	Description
NPN		When base is high current flows from Collector to the Emitter. **N**ever **P**oints i**N**
PNP		When base is low current flows from Emitter to Collector. **P**oints i**N** **P**ermanently

The base is actually "the switch" or lever that is releasing the current from our collector to the emitter, in the case of a NPN. The greater the current that we supply to the base, the more current will be released from the collector to the emitter. How much will the NPN or PNP multiply the current from the collector to the emitter? It depends on the part. The BC547 NPN transistor is described in the data-sheet as "general purpose application" and "switching application".

This part will still multiply the current but not as much as perhaps another NPN might have done. The data-sheet provides a graph similar to the one below.

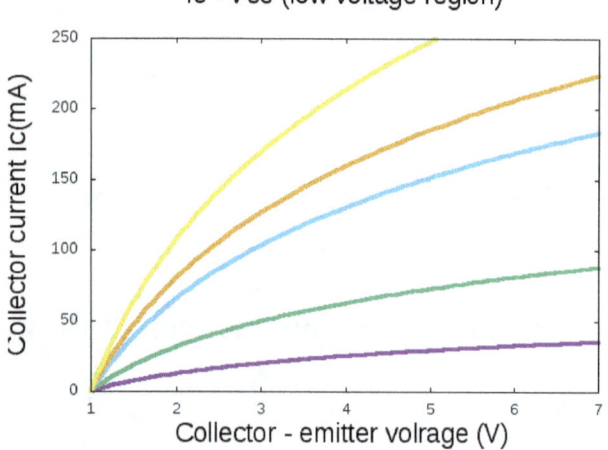

It is possible to see how our input current is being multiplied into a larger current at particular voltages. One example is that .2mA of current at 3 volts will be virtually 40mA on the emitter.

Short circuits

> "If it doesn't burn it is software"
> Mikhail Stoliarov

A short circuit is an abnormal connection between two elements of an electrical circuit bypassing the rest of the circuit. In some cases the voltage is different than the intended voltage due to little or no resistance. One of the side effects is the rapid built up of heat.

This is a nice way of saying you have made a mistake. Something was connected that should not have been connected or without enough resistance to protect one or more components. Although it may not be obvious, it is usually the heat at the source of the short circuit that causes the components to fail. The most obvious example is that of a LED fed by too much current. One of the side effects of a LED is the light but when fed too much power it is not able to dissipate the heat fast enough and that will eventually causes the LED to fail.

Software can also have problems that cause things to go awry, yet other than a few wasted CPU cycles there is usually no damage[13]. With hardware, failure to verify connections will be the fast track to creating a small lump of no longer functional plastic.

There is no magic to prevent this from happening. Fuses or circuit breakers can be used in

---

13 Not entirely accurate but go with me on this.

electronics just like in cars and houses to help protect the hardware from overloads. Other steps that can be taken are more procedural such as not having metal things lying on your working area and double checking your wiring before powering up your work.

In software systems it is possible to debug them by adding print statements or logging information as the program runs. With hardware this is not possible, but it is possible to use your multimeter to verify that the voltage is at the values you expect at various points on your circuit.

When soldering it is not only possible to verify your resistors are properly soldered, but also good practice. This is less of an issue for normal resistors but a real concern when soldering on SMD parts. All evidence to the contrary, not all short circuits are self made. There have been plenty of instances where I was too casual with the number of metal things laying on my desk, or when plugging something on a breadboard in haste.

Yet the most embarrassing problem was when testing out a new piece of hardware, a digital temperature sensor. I didn't look at every single letter on the diagram. It turns out that I should have. The diagram for this sensor was incorrect at worse or very misleading in the best case. The label for power and ground were incorrectly labeled.

Illustration 14: Incorrect schematic

If you chose the incorrect setup when wiring it up, you may never have a second chance. It turned out that the side connectors were correctly wired and described on the board but also incorrectly described on the forms that came with the board.

It may not be obvious but on the schematic diagram seen in Illustration 14 you can see that SCL, SDA, Ground and 5V is the order but on the actual PCB the order is SCL, SDA, 5V and ground.

2 TOOLS

If you are planning to solder together a kit or even just a few parts then a soldering iron, solder, a wire cutter and a pair of safety glasses is all that is really necessary. Right? In theory all that is necessary is the ability to heat up the wires enough to apply solder to them. In most cases this minimum set of tools is perhaps too spartan and a few more tools would help in the creation of projects. There is literally no limit to how many tools you can have but I will describe a nice subset that will prove helpful in most cases. A list of these tools and a brief description of each and their uses follows.

Safety First
A pair of safety glasses is a good idea in most any situation. When soldering you are dealing with liquid metal at a temperature of least 183 degrees Celsius. This temperature can cause 2nd or 3rd degree burns on hands or skin but would be devastating if it got into your eye. Use eye protection when soldering. The best would be a pair of safety glasses but anything between your eye and the soldering iron is better than nothing.

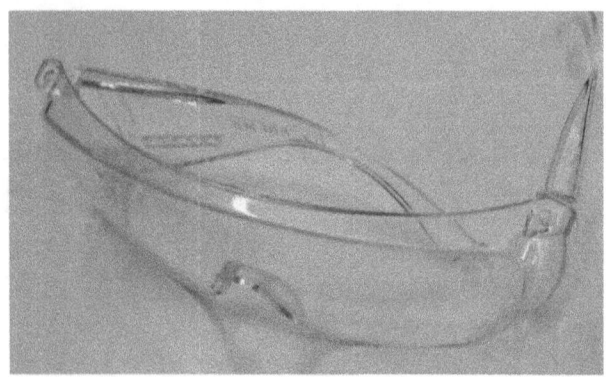

Illustration 15: Safety glasses   Illustration 16: Head mounted magnifier

Fume Extractor
Work in a well ventilated areas. Lead will vaporize at 485 degrees Celsius which is considerably hotter than the recommended 300 degrees Celsius you will be using so it shouldn't be a concern that the lead is vaporizing.

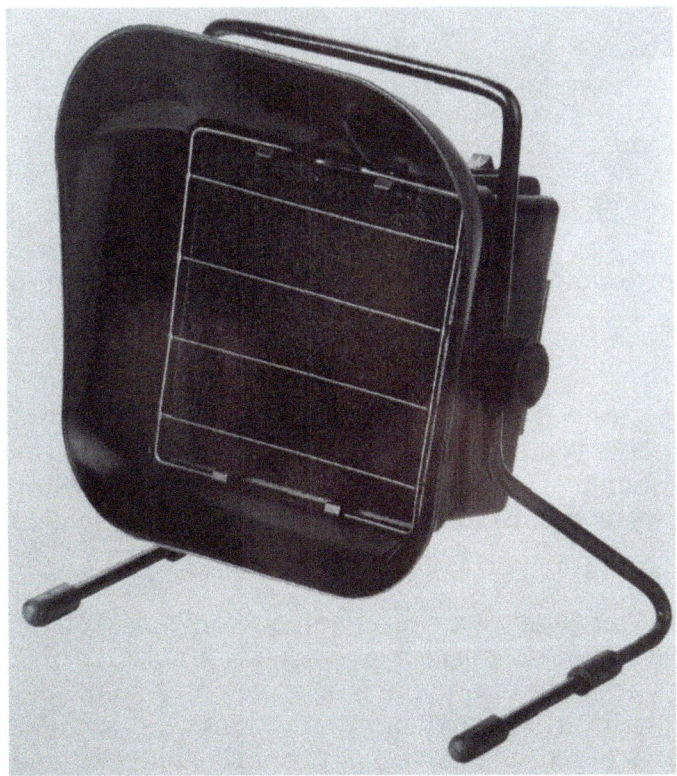

Illustration 17: Fume extractor

However, the flux that is in your solder will give off gases while soldering. This is just as true for soldering as for other types of activities that include glues, paints or solvents. These gases can cause irritation and asthma like symptoms.

The mask
This mask won't help one tiny little bit when dealing with toxic gases or vapors but will help to keep other small particles out of your airways.

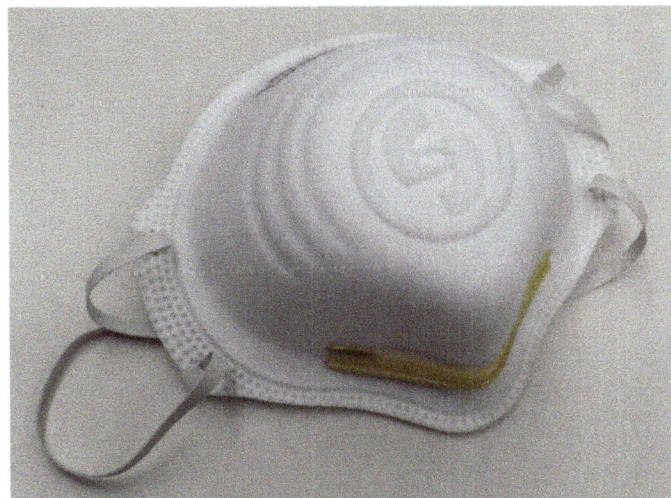

Illustration 18: Medical mask

Wearing such a mask is a good idea when doing grinding or when cutting will produce fine particles.

A clean desk ...
... is the sign of a clean desk and parts laying on your working area can cause short circuits to your project as well as turning them into an unexpected sources of electricity.

This is not a definitive list of all the safety hazards or safety equipment you could encounter but simple items can be used to give you more control over your environment. This can be as simple as a power strip with an on/off button or a standard fan for helping with ventilation.

Multimeter
A multimeter is an electronic measuring tool which combines several different measurement functions in one instrument. The typical multimeter's functions are the measurement of voltage, current and resistance. This device is the final word when measuring your resistors. This is important if your color chart is not at hand, but it is also convenient for measuring unknown parts.

This is one of the few debugging tools of the hardware world. It is not only useful for individual parts but vital to determine if the voltage is correct at various junctions of a circuit, if there is a short circuit, or to simply verify that your parts are properly soldered. These types of verification's are more obvious with larger parts but as the parts get smaller and smaller then the naked eye cannot easily discern if everything is properly soldered and that is when the multimeter really shines.

Illustration 19: High end multimeter

Illustration 20: Regular multimeter

All multimeter will measure resistance, direct current and alternating current. Yet some of the more advanced models can measure capacitors or can even test transistors. These features are nice to have but are not critical when deciding on which multimeter to purchase.

Illustration 20 and Illustration 19 are two different multi-meters and at first glance the one with more settings appears to be the better device, and actually it does offer more options but does that mean it is best? Should you buy it? When doing a measurement with this high end multimeter you need to know which range your measurement will fall into and select that setting accordingly (e.g. 2K ohms to 20K ohms).

The simpler multimeter frees you from this step. When testing resistance you simply choose a setting(e.g. resistance) and test the component, and the device will determine the range for you and then will display how many ohms the resistor has. Depending on your needs it might be much more convenient rather than fiddling with each of the ranges in order to measure your components.

This symbol on my device is for the continuity test. Simply put this determines if there is an electrical connection between two points. Although this may sound perfect don't forget that this test will not work through some components or in some situations. One such situation is you will not be able test a connection through a resistor.

This setting tests the resistance of an object. Depending on the functionality of your multimeter, illustration 20, start off at the lowest setting and touch each end of your resistor with one of the probes. You will either get a value and thus are satisfied, or you will probably get the value one. In this case it means that the value is off the scale, simply advance to the next range and test again.

If your multimeter does not have different ranges you simply touch your board or component to measure the resistance. After a brief time, the multimeter will determine the resistance and display it on the screen.

You can measure the DC voltage in a similar manner as measuring the resistance.

You can measure the DC voltage in a similar manner as measuring the resistance.

Measuring the actual current for either AC or DC is also possible but this will require a slightly different setup. Unlike the voltage, if you want to see how much current is being used you cannot simply touch different parts of your circuit with the probes. In order to measure the current, you will need to actually insert the multimeter into the circuit. This is quite possible even so it is a bit invasive to start cutting wires in order to add your multimeter into the circuit.

It is difficult to know in advance if there will be a problem with a circuit but if you are designing your own circuit boards then it would be forward thinking to put in a pad or a jumper in locations that have proven problematic in the past or might be near difficult items to solder.

Probes

The probes are at the end of the two cables which are connected to the multimeter. This allows the user to select which two points in the circuit should be tested. The multimeter will have several plugins which the cables can be connected to. Which one you connect them to depends on the desired functionality. Due to technical reasons the multimeter's internal circuit has mutually exclusive logic internally and thus it is necessary to decide which set of measurements you want to take by plugging in the probe into the appropriate plug.

Different multimeters will support different functionality and may have a different number of plugins but the one common thing is that each device will have is a COM (common) plug. This plug is used for ground no matter what other choice is selected.

The multimeter can be an amazing companion in diagnosing problems or can be used for some simple measurements. Read your user manual to get the most from your device.

Note: You have to change your setup to use the amperage measurements and you should be careful as feeding the wrong voltage to the wrong setting will at best give no results and at worst could damage the device.

Note: This very small guide is meant to be a brief overview, not a replacement for the manual that came with your multimeter. The Internet[14] is a valuable resource for seeing other people's viewpoint or experience on a given topic. The internet can and should be used to gather more information on a topic, but only after you have a basic understanding.

Wire Cutters

For most situations a standard pair of wire cutters will be fine for cutting wires or clipping short component leads. Yet, if you have a little bit of free budget you might want to consider the purchase a set of flush cutters, see Illustration 21.

---

14 A very nice basic tutorial on the multimeter. http://www.youtube.com/watch?v=bF3OyQ3HwfU&feature=fvwp&NR=1

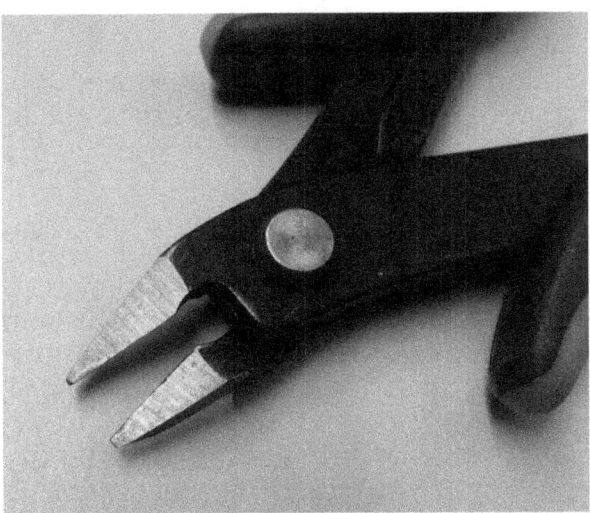

Illustration 21: Flushcut wire cutter

Flush cutters also can be used to clip off small leads after the component has been soldered to the board. However, the reason flush cutters exist is because the blade has been angled in such a way that you can clip very close to the soldered joint. These angled cutters will leave either a very short or non-existent stub on the board. This is especially important if the board itself will not be in a case.

Tweezers
A pair of tweezers will be necessary but which pair is best suited for you is determined by the soldering work that you will be doing. No matter what types of project you do your tweezers will probably not be the same as those that you have in your medicine chest.

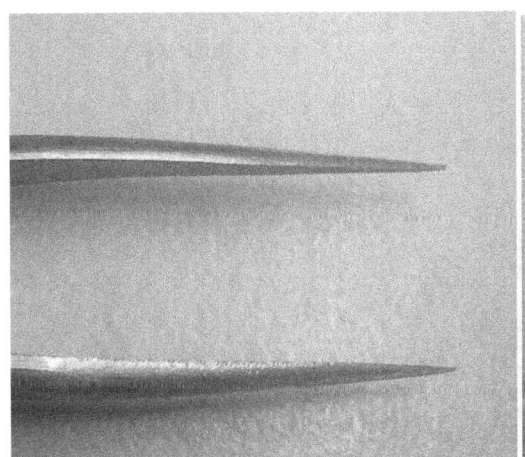
Illustration 22: Best tweezers

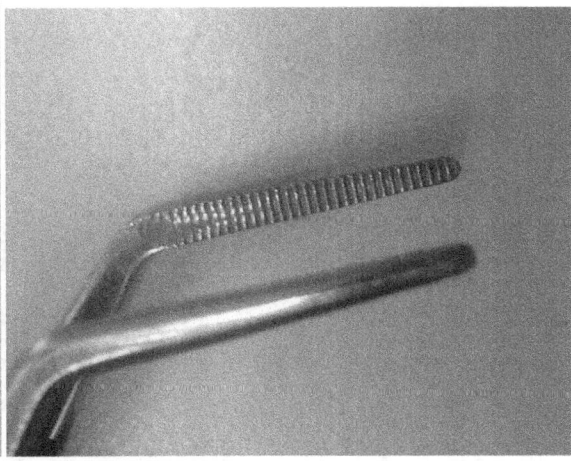
Illustration 23: Lower quality

Small surface mount components will be easier to hold using a pair of pointy tweezers. It is easier to both solder your component while simultaneously keeping the tweezers out of your own way. If you do any work with surface mount parts you will find very pointy tweezers to be not a "nice to have" but an actual "must have".

Yet that is not to say that a fairly standard set of tweezers may not show their worth specifically because they are large. Wires will conduct electricity but because they also conduct heat this may create some problems when soldering. This problem is easily seen when trying to solder a very short wire because the heat is transferred down the wire to the next solder joint which is potentially undoing you efforts.

A large pair of tweezers will act as a heat sink when put between the soldering iron and the rest of the circuit. In addition to holding the wire steady it will direct the heat towards itself rather than passing it further down the wire to other soldered parts or joints. The tweezers will also give you some control over what you are doing while redirecting the heat.

Another use for large tweezers may be when de-soldering parts from a circuit board. They will give you a good grip on the part to be removed while using the soldering iron.

Third hand
It is a very flexible person who can hold the components to be soldered, the board, the soldering iron and the solder with only two hands. There is a tool, called a third hand or helping hand, Illustration 24, which can be used to assist in this task. The third hand usually has a alligator clips at the end of each arm and sometimes there is also a magnifying glass attached. The board and components can be held by the third hand thus freeing up your hands to hold the soldering iron and solder.

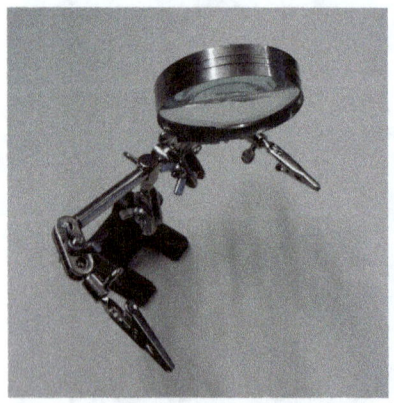

Illustration 24: Helping hand

Illustration 25: Board holder

I have also discovered another tool which despite its shape, could very well also be classified as a form of third hand. This tool, Illustration 25, is not designed for the same degree of flexibility as the third hand. It is designed to hold a printed circuit board. This can be used for through hole soldering but it excels at holding a board for surface mount soldering.

Homemade Tools
A homemade tool could be anything from a pointy stick you use in the garden to a self built computer designed for a single purpose. It is quite likely over time you will add to your collection some homemade tools.

A stick
Well, not exactly a stick but rather a piece of PCB board which was used as a spacer when soldering resistors to the board. It was also used as a spacer when trying to build a LED cube.

Illustration 26: Piece of circuit board

This tool has the advantage that it doesn't conduct heat, is thin and won't burn.

GPIO Header breakout cable
The fact that the Raspberry Pi has a header that can be conveniently accessed is great but after you attach a couple of wires it can become more problematic to verify that you are using the correct pin. You cannot make more space on the pi but you can "move" the connections so they are easier to have a good view.

Illustration 27: Homemade breakout cable

This was built using a standard 26 pin flat cable with a female header, Illustration 27, on one end and even those pins are connected to a project board which had some headers soldered to it. The headers are using the standard 0.1 inch spacing (.254 cm) and so it can be plugged directly into a breadboard. This is a common problem when the Raspberry Pi first came out and now possible to find similar breakout boards for sale.

Mosfet Cable

During a board development I wanted to use a SMD format of a specific mosfet. However I did not want to use this part without first testing it. Because this was a surface mount chip it would not have been possible to test on a breadboard.

Illustration 28: Mosfet tester cable

In order to test it, I simply soldered wires to it and used shrink wrap to help ensure that the connections were solid. Once this was done the chip then had proper wires so this allowed us to test on our breadboard and verify if this component was acceptable for our circuit. This was not so much of a general tool as much as a simple piece of test equipment.

LED and Wire

This is a simple wire with a resistor, a LED and a female connector so it can be connected directly to the Raspberry Pi's GPIO pins directly during testing.

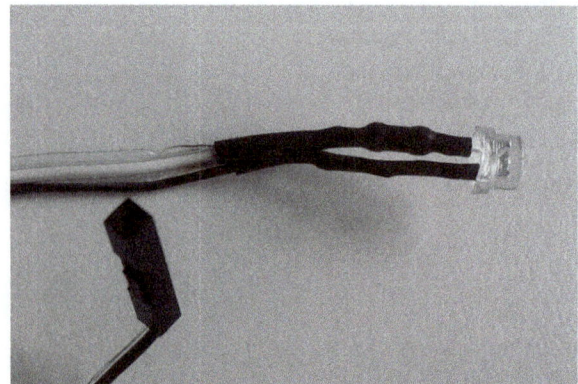

Illustration 29: Wire with LED and inline resistor

This also was not so much of a general tool as much as a simple piece of test equipment. The problem occurred often enough that we needed to connect up a LED to a set of header pins. The effort to create such a tester was so low we created this part.

This tester was created using a purchased wire that had an end already designed for connecting to a header pin. A resistor was then soldered inline with the LED so this would never be forgotten as part of the test setup.

A wire and resistor
It is fairly often that I need to test a LED or test a solder joint for a LED. This was especially often during my creation of the LED cube. Even though my first cube was only a 4x4x4 it did contain over a hundred solder joints.

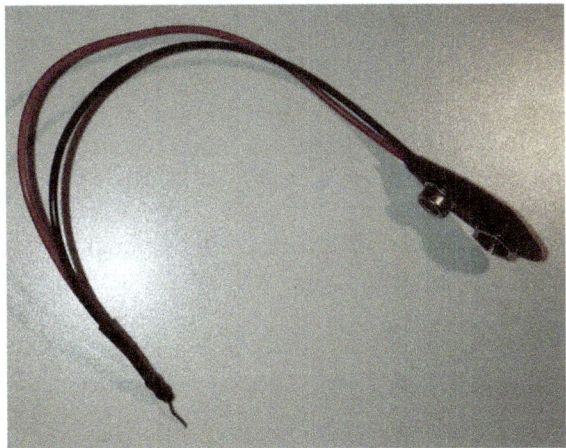
Illustration 30: 9V battery connector with inline resistor

Just like the other self made tool I had to add a resistor to the wire in order to not burn out the LED.

Soldering iron
It is possible to purchase a simple 15W soldering iron similar to the one in Illustration 31 or yet a better choice might be a 40-50W iron, Illustration 32 with an adjustable temperature.

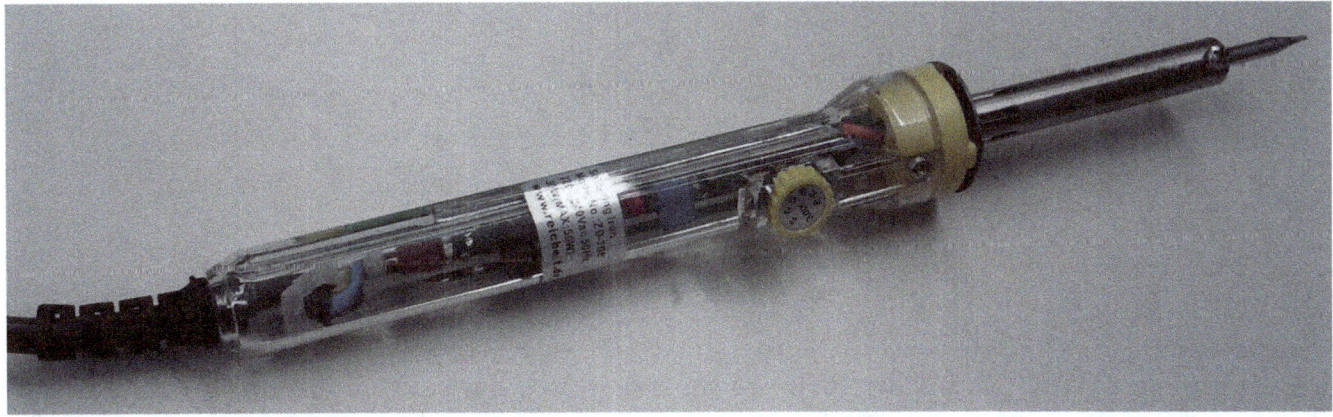
Illustration 31: Pen style soldering iron

A simple soldering iron will melt the solder just as well as fancy one but the extra power would be nice if you need to de-solder parts from an electronics board. Consumer electronics tend to use leadfree solder which does have a higher melting point.

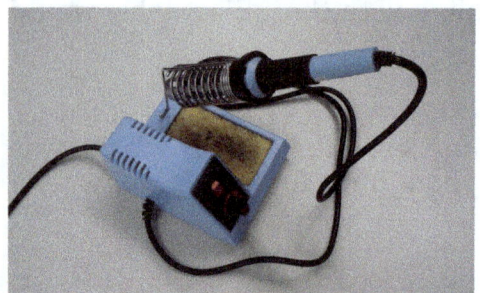

Illustration 32: Soldering station

A second advantage to a soldering station is that they usually have replaceable tips. Most times a new solder iron will have a conical tip yet this is not the only type of tips available as seen in Illustration 33. Some soldering irons come with a chisel tip which is so named due to its appearance. Each different style tip tends to be used in a different situation. This is why so many styles are available.

In addition to the types of tips there can also be varying sizes (e.g small conical versus large conical).

- Bevel
- Chisel
- Large or small conical
- Concave (not shown)
- Knife edge (not shown)

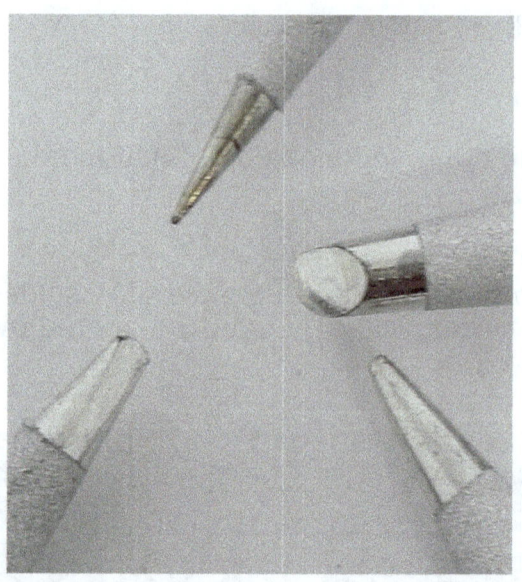

Illustration 33: Multiple styles of soldering tips

It is important that the tip be an appropriate size for the components you will be soldering. Depending on what types of parts you are soldering, you may wish a larger or a smaller tip. If your project contains many small parts it might be intuitive to think a small tip would be better. After all, a small tip will allow you to get in close to the component you are soldering. Although true, a small tip can also make it difficult to solder some components. Small tips can actually make it harder to do the heat transfer to larger components or components with a large heat sink.

In general a conical tip may be good for soldering quite small or surface mount parts while the chisel tip may be better for larger parts or for de-soldering parts. However, there are many people who swear the chisel tip is just as good for soldering surface mount or small parts. In the end this all comes down to personal preferences. A good soldering iron would be one that has a number of different tips or has the possibility adding new tips in the future. At the very least you can replace tips that have worn out.

A soldering iron stand

If you purchased a soldering station with adjustable temperature it probably already came with a stand to hold the soldering iron. If not already provided, it is a very good idea to purchase a good and stable stand. This point cannot be emphasized enough, a simple soldering iron might be sold with no stand or a very small stand which is designed only to elevate the soldering iron tip from the table.

Illustration 34: Soldering iron holder

This simple soldering iron stand, Illustration 34, is sold separately and is designed to hold the iron securely away from the workbench. This is much better than nothing but this stand does have the drawback of being rather light. This could be improved by attaching some weight to it's bottom.

It should be obvious why a stable stand is important – the soldering iron is hot. In most cases it will be several hundred degrees which can melt or mark up surfaces, wires, or hands.

Soldering Basics

Soldering is joining multiple metal items together by melting a low temperature metal alloy(solder) and having it flow in and around the items to be soldered. Soldering joins components to each other or a printed circuit board forming not just electrical connection but a sturdy connection.

Sometimes people confuse soldering with welding. The results may look similar in some circumstances but the technique is completely different. Welding is the act of melting two pieces of metal that once it cools down will form a strong material bond. This bond cannot be reversed or undone. Soldering on the other hand, is adding a metal alloy to an electrical connection which when cooled provides a solid electrical connection.

Unlike welding it is possible to more or less reverse a soldered connection. This is possible because the two parts are surrounded by solder. Because the solder melts at a lower temperature than the wires it connects it is possible to reheat the solder and then carefully remove the part or wire from the joint.

Solder

Solder is a metal alloy comprised of several different metals which has a relatively low melting temp. Leaded solder will melt at about 285 deg Celsius while copper will melt at 1100 deg Celsius.

In addition to selecting the right soldering iron, you need to select the right solder for the task at hand. One of the first things to consider is the diameter of the solder you are using. The reason for this is that when soldering it is difficult to control how much solder flows onto the joint when the solder wire is too thick. Of course thicker solder (1mm) can be helpful when tinning wires, but the smaller the components get, the greater the need for a smaller diameter of solder.

The thickness of the solder is measured in millimeters. A nice thickness of solder for project work is 0.5mm. This is still big enough when soldering through hole parts, but small enough that you can solder SMD parts if you want to.

Solder Composition

In the electronics industry alone there are many different types of solder, Table 9, that can be used to join together components. The different types are used in different situations by different groups of people.

Industry tends to use lead free solders while most hobbyist use leaded solders. The reasons for this is both because of the types of equipment as well as differing abilities. Leaded solder has a lower melting point and thus should make it easier for the average person to solder components together without applying too much heat and damaging them.

Composition[15]	Composition attributes
$60_{sn}\ 40_{pb}$	Quite common and easy to use
	Can create cold joints if part is moved before it cools completely.
	Melts over a small range of temperatures (183 C – 188 C)
$63_{sn}\ 37_{pb}$	Quite common and easy to use
	Melts at a single temperature (183 C)
Lead free	Hotter iron is needed.
$96.3_{SN}\ 0.7_{CU}\ 0.3_{AG}$	Higher risk of component damage due to hotter iron.
	Higher melting temperature (217 C – 221 C)
Lead-bismuth eutectic[16]	This is a special mixture of Bismuth and Lead. What makes this
$55.5_{bi}\ 44.5_{pb}$	special is that it both has a lower melting temperature and it cools much slower. These attributes make it ideal when trying

15 Who ever said the periodic table wouldn't be useful in later life. SN = Tin, PB = Lead, AG = Silver

16 Eutectic mixture is one that melts or freezes at a temperature lower than any of the components.

	to de-solder components. The slow cooling makes it possible to heat both sides of a big part and then easily pick it off the board before the solder cools. Melting temperature (123.5 C)
Solder paste	Solder paste much like the name implies is solder that is applied in a paste like form. It is a mixture of the powdered solder in a gel rosin creating a paste like mixture that is primarily used for soldering surface mount components.

Table *9: Different types of solders and their attributes*

Lead Free solders would not be the best choice when first learning to solder due to it having a higher melting temperature. It is recommended that you start off first with a standard leaded solder and once the techniques have been mastered, then if necessary then try out soldering with lead free solder. If you do advance to using lead free solders you may also need to purchase new tips for your soldering iron as the tips are plated differently depending on the type of solder being used.

As of July 1st 2006 all electronics in Europe must be made with lead free solder. This was part of two pieces of legislation, the Restriction of Hazardous Substances (RoHS), and Waste Electrical and Electronic Equipment (WEEE). However, It is still possible for hobbyists to purchase leaded solder in some of these countries. At the present time, there are no laws preventing the use of leaded solder in electronics in the United States.

Note: Which is the best composition depends on the actual circumstances, but for the home hobbyist the 63/37 is both very common and probably the best choice.

Flux
Flux is a component in virtually every different solder mixture in order to perform two different actions. The first is to clean oxidation from the surface of the components that are being soldered together. The surface may have a small bit of grease or oxidation on the pads or the wires being joined. The flux also helps with the wetting. Wetting is a condition where the solder has become molten at its eutectic temperature with an adequate amount of flux. The solder in its molten form will then adhere to the components or the pad on the circuit board.

Just as there are different compositions of solders, there are also different types flux available. One of the most common flux's in solder is rosin[17]. The two different types of rosin flux is Rosin Active (RA) and Rosin Mildly Active (RMA). The RA is more aggressive when removing impurities and oxidation, but the two are similar enough for hobbyist use.

Organic Flux is interesting because it is water soluble which is helpful for cleanup. The downside is that this flux will continue working unless it is cleaned off the boards and joints effectively. If all of the organic flux has not been cleaned from the board it can later react with moisture and cause unexpected electric conductivity which may result in short circuits or other board failures. Organic type of flux has more downsides than upsides for the hobbyist.

One thing not yet mentioned is that most 63/37 leaded solder has a rosin core. This ensures that some flux will be in the solder when soldering. Yet flux does burn off over time

---

17 Rosin is a form of resin acids which are extracted from pine trees and some plants.

which is why the soldering iron tip needs to be periodically cleaned.

It is possible to separately purchase or even make your own flux. This is very useful for cleaning oxidation off the pads on a board. A little extra flux is quite helpful when soldering surface mount chips that have both many legs that are close to each other (e.g. 0.7mm separation)

The act of soldering
During most of the book, the projects will be either configuring the Raspberry Pi, Arduino or working on circuits using a breadboard. This works well for experiments but breadboards are more suited for the workbench and may not be sturdy enough nor small enough for embedding into projects. Even if they are small enough the breadboard won't be stable enough for any place other than a clean desk.

Despite how fragile even the smallest gauge wires are, once they are soldered onto chips, LEDs, or project boards they do form very solid and stable connections.

Soldering techniques
Soldering new components together is actually quite easy especially if you have a steady hand. Soldering is simply heating up the parts to be fused together until they are warm enough that the solder can melt on and around them.

The technique of soldering is to put the tip of the iron behind the wire or lead and on the pad to be soldered so it can heat up. Once the metal is at the melting point of the solder, pressing the solder onto this joint will cause it to melt and flow around the wire or component. It takes practice to know how long to heat the part before applying the solder, but if your iron[18] is already heated then this might take between two and three seconds.

Electronic components can be exposed to these high temperatures while soldering but usually only for a short period of time. The longer the item is heated the more likely they will be damaged by the heat from the soldering iron.

When doing through hole soldering you can bend the leads slightly after inserting them through your board. This will create a small amount of tension and holds the part in place so it doesn't fall out or move around when soldering.

It might be tempting to melt a small amount of solder on the tip of the soldering iron and bring that to the wire or component to be soldered – don't do it.

By melting the solder onto the tip of your iron you will be burning off the flux. The flux is helping to ensure that any residues or oxides are not on your joint, and it is this that allows the solder to flow. Dragging liquid solder over the joint may reluctantly stick to the wire but the result will be blobby, dull looking or grainy appearance. These are the traits of a cold joint. A cold joint is brittle and is prone to fail at some point in the future. If this connection doesn't fail immediately it will be likely to fail in the future.

You do not need to clean the tip of your soldering iron after soldering each part but it will need to be cleaned periodically. This is because it will slowly connect more and more old solder which will make the task of soldering more difficult. The reason it is harder to solder is because the flux has been burned off which increases the chances for creating cold joints.

---

18 The recommended heat for soldering is 330 – 350 Celsius

Note: Never flick solder off of your soldering iron always carefully wipe it on your moist sponge or brass sponge.

As simple as this sounds, this is most everything you need to know in order to solder parts together. Although practice makes perfect, there is one more bit of organizational advice that should be considered. That is to place parts starting in the middle of the board and work your way to the edges. Always start with the smallest parts and do the larger parts last.

It seems somewhat intuitive when said so succinctly but you must remember that in almost every case the largest item will be your hands and the soldering iron. Careful forethought on which components should be soldered first must be done so these components do not complicate soldering the rest of the parts on the board.

Always wash your hands when you are finished. Working with leaded solder, electronic components of known or dubious origin will undoubtedly leave substances on your hands that you would not knowingly ingest.

Lessons learned
It cannot be stressed enough that patience may be one of the most important skills being used when soldering.

Just like any major undertaking planning is important and just a few seconds or even a few minutes setup before soldering a part will pay off in the long run. The integrated circuits, LEDs or other connectors can indeed handle high temperature for a brief time but too much heat can affect not only the component but perhaps the electronic board itself.

A good example, or perhaps a bad example, is an attempt of mine to solder a microphone jack. I was too hasty and when the metal connector was still hot, it was slightly pulled and it ruined the part. It might be difficult to see but if you look at the marked area on Illustration 35 you will see some white substance on the connector. This is most likely the plastic that was holding the tiny metal plate into the audio connector.

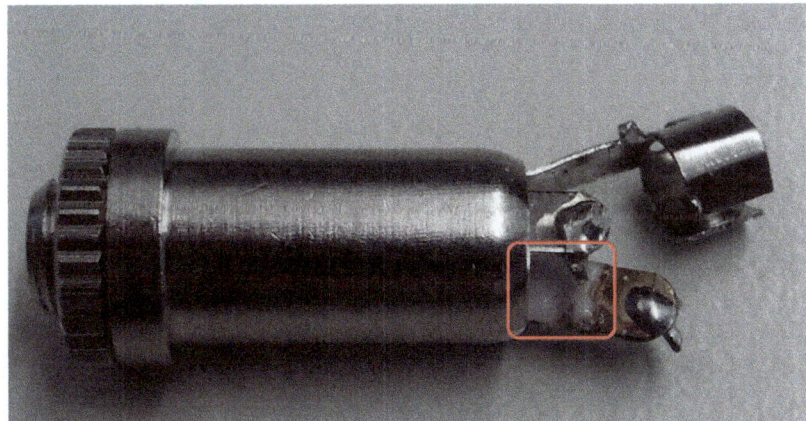

Illustration 35: Example of a ruined part

Excess heat
The electronic parts are built to withstand some heat during the soldering process. How sensitive the parts are to the heat during this process depends on the part and its size. The

heat is transferred through the part itself, but this effect doesn't need to be limited to electronic parts. The same effect also takes place through wires connecting parts or on the circuit board itself.

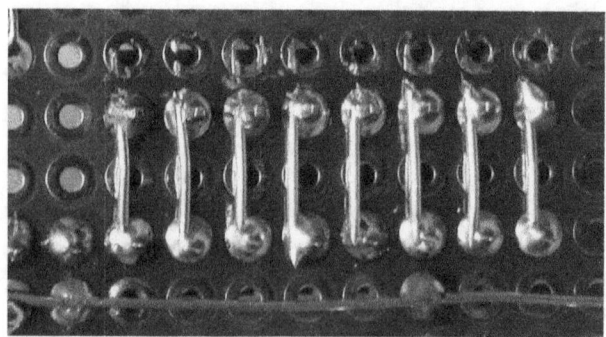

Illustration 36: Situation where heat transfer can create soldering problems

In this example, the wires were very short and it was not possible to solder one end of the wire without the heat being transferred and melting the joint on the other end of the wire. The solution is to use some sort of heat sink that will conduct the heat away before it goes to a sensitive part or in this case to the other solder joint.

For this particular example, I simply held the wire in place with a very thick pair of tweezers which acted like a heat sink and allowed both ends to be soldered separately. Soldering two or three legged components have the possibility of heat damage if the soldering iron is left on too long. However, when soldering many legged IC's there is an even greater chance of this happening.

To mitigate this situation there are a couple of things you can do such as soldering every other leg on the chip. This will help to prevent the heat from building up internally. Another approach is to alternate between pins on one end of the chip, with pins on the other side and other end of the chip. Finally, you can also simply solder a few pins and take a small break. The idea is to give the heat a chance to dissipate.

Soldering sockets
Without much practice it is possible to solder a few wires to each other, and it is even easier to solder a wire or component onto a circuit board. Simply place a hot soldering iron onto the pad and wire and press on the solder.

When assembling your own projects you may opt for a socket to hold your IC, especially if it cost more than a couple of dollars for the single part. Who wants to discover they overheated their micro-controller while soldering and ruined the part so you must then start again with a new part. The easy solution is to solder socket and insert your chip when you are all finished.

External wires
Soldering on extra wires can create new challenges. The problem usually has less to do with soldering and more to do with space. It is virtually impossible to solder two wires to that small leg that poking through your pcb board without a bit of preparation.

Simply solder the socket onto the board as you would any normal part or wire. To connect a wire or lead to that socket, trim the wire to the correct length so it will terminate at the connector, lay the wire on top of the joint and briefly press on it with a hot soldering iron.

Illustration 37: Step 1, Place wire on the solder joint

Illustration 38: Step 2, Press wire into solder joint

The soldering iron will heat the wire, melt the solder and cause the wire to sink into the joint. You may need to apply a small additional bit of solder while performing this.

De-soldering Basics
Eventually you will come to a situation where you need to remove something that you have soldered to a board. This can be done using a couple of different methods. The first is to simply heat the component or joint with the soldering iron and carefully pull it off the board. You must be careful when doing this otherwise you will pull the pad off the board as well.

Another situation would be if too much solder ended up on the board or joint. In this case the whole component doesn't need to be removed but rather just some of the excess solder. Solder wick from Illustration 39 is a braided metal band that contains flux. Simply place this on the area with too much solder and heat it with the soldering iron. The heat and the flux from the wick will cause the excess solder to flow into the wick and off the part.

Illustration 39: De-soldering wick

A tool specifically made for assisting with de-soldering parts is the de-soldering pump from Illustration 40. There are several different styles of pump, the plunger style pump which is a cylinder with a plunger that pushes down a small spring loaded seal that is locked at the bottom near the opening. When the user presses the button on the side will release the spring and then the seal which is pulled up and creates a small vacuum effect. This is placed over a joint where the solder is fluid so it can then be sucked into the cylinder and off the part and board. Removing excess solder in this way does require some hand/eye coordination.

Another style de-soldering pump is the bulb style. This looks a bit like a very short turkey baster where you press the bulb and place it next to the fluid solder. When you release the bulb the air and hot solder will rush into the pump in a similar manner to the plunger style pump.

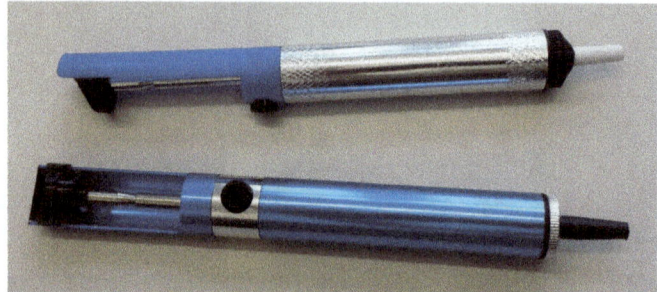
Illustration 40: Solder sucker

Don't leave the soldering cylinder cocked when storing it as that will stress the spring and probably shorten the life of this tool.

It may not be intuitively obvious but when de-soldering you may need to use a small bit of solder. Having a just a tiny amount of solder on the tip of the soldering iron may be necessary while de-soldering. This small pool of liquid solder will help in transferring the heat to the cold joint that you wish to de-solder.

One thing to watch out for when de-soldering the parts, especially from other printed circuit boards, is that the leads may be bent slightly out. When de-soldering the part the leg will still be bent out and it may even be touching the inside of the hole. You may need to push at the pin slightly with either a pair of tweezers or perhaps even the tip of your soldering iron to ensure the part is not being held either by tension or still soldered to the side of the through hole.

When trying to de-solder other slightly larger pieces from the board it may be tricky depending on how many legs there are on the IC or socket. You may need to use a small flat screwdriver to help get the socket off the board, but be careful. **If the socket or IC is still soldered to the pad, pulling it up especially while heated may cause the pad to come off the board.** If the part is defective or inexpensive it may be cheaper and easier in the long run to use your flush cutter to cut off the component and simply de-solder the leg that is left on the board.

Hints after de-soldering
It may not be enough to de-solder the part or to clean the solder off the board with solder wick, there may still be some solder in the hole. Depending on your part this may be ok, the resistor has a very small wire that may still go through, but this may not always be the case.

One thing to remember is to use a toothpick. It is round and will not stick to the edges when the solder cools down. Simply touch your iron to the pad and heat it up and press a toothpick through. This will give you a hole that may be almost as good as new, it should again be round and any extra solder will be either pushed to the edge or to the other side. In any case, this should most certainly be better than what you started with.

Advanced soldering techniques

Over time electronics have moved from vacuum tubes to printed circuit boards. Circuit boards initially used only through hole components. Through hole components are components with wire leads going through the PCB and are then soldered to it. The overall size of the components were reasonably large in size which limits the density of the components which in turn limits the size of the board.

Surface mount components started to become popular in the 1980's. Surface mount components are much smaller and are placed directly on copper pads on a single side of the PCB. High density of component placements on a single side means that printed circuit boards can be smaller. Unlike through hole boards, surface mount boards are smaller and both sides of the board can contain surface mount parts.

It is possible to use a standard soldering iron and regular 63/37 solder to solder these components. This does get harder and harder as the size of the components shrink. To counter this different techniques exist for soldering parts to printed circuit boards. It is possible to solder these parts with a standard pen style soldering iron but the size of the tip usually dwarfs the part which makes it challenging to get a good solder connection. The solution for this problem was solder paste.

Solder paste much like the name implies is solder that is applied in a paste like form. It is a mixture of the powdered solder in a gel rosin creating a paste like mixture that can be used for soldering through hole components but is usually used with surface mount components.

Solder paste is smeared on the pad, the component is set on top of it and then heat is applied. This heat can be a hot soldering iron that is briefly touched to the pad, leg or component causing the solder paste to melt and the solder to flow. The result is that the part is soldered to the board.

It would still be quite time consuming to carefully squeeze out small amounts of solder paste on a printed circuit board that has many pads. The solution is to use a soldering mask when applying the solder. A soldering mask is essentially a piece of material that covers the PCB but has holes where each pad is on the circuit board. Then solder paste is applied at one edge of the stencil and a device similar to a flat squeegee pulls the paste over the stencil where it is goes through the stencil and is deposited on the board. When the soldering stencil is removed all that is left is a very thin amount of solder paste on all the individual pads.

It would also be time consuming and possibly error prone to use a soldering iron next to each individual surface mount component when soldering them to the board. The industrial manufactures came up with a solution that was developed to deal with this problem. As silly as it sounds the solution was essentially to place the printed circuit boards containing both the solder paste and the components into an oven – a reflow oven.

The reflow oven is used to apply heat to the solder paste without directly touching the component itself. The air is circulated until both the solder and the adjoining surfaces of the surface mount device and pad are heated. The solder becomes liquid and the process of soldering is complete. One rather neat effect of soldering parts in this manner is that the flux and molten solder causes the part to float on the solder and then snap to the proper orientation on the pad.

A second way of heating solder paste is by using a heat gun. A heat gun also produces hot air that can be directed at a fairly small spot. This could be used for surface mount soldering but it is more often used in repairs. The practitioner will use the heat gun for heating up a component that needs to be removed and once the solder is heated so it is once again fluid the old part can be removed.

Final word on soldering
The process of soldering is quite easy and can be mastered with a small amount of practice. The road to mastery comes not only with practice but also with research. This can come in the form of tutorials and videos which is available in great quantities on the internet. One of my favorites is a very readable guide in the form of a comic book by Mitch Altman, Andie Nordgren and Jeff Keyzer.

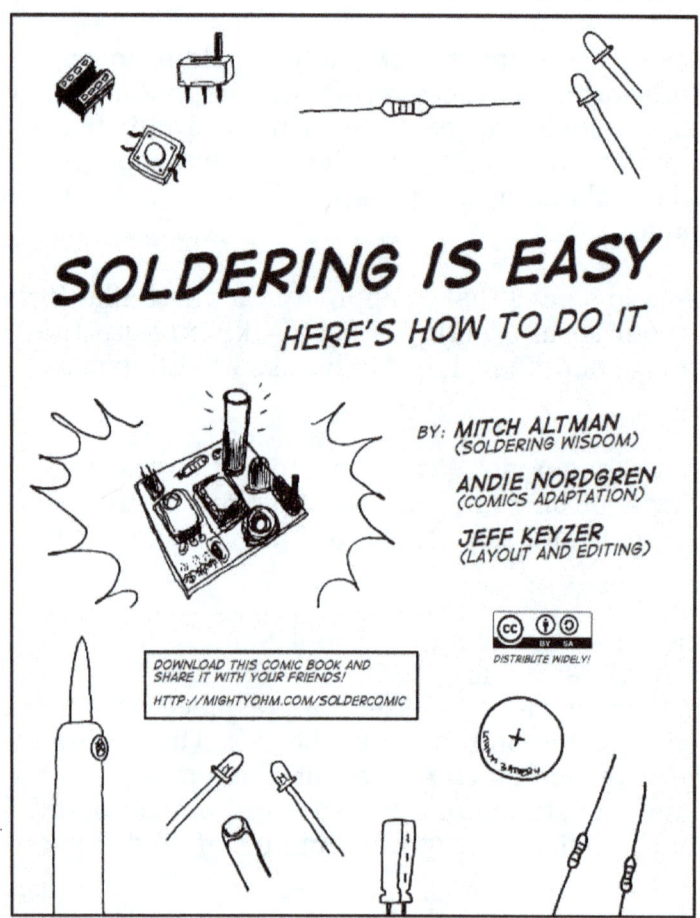

This guide displays the essence of how to solder through hole parts with big friendly pictures and the minimum amount of words. This is not just a casual project to spread the word on soldering but a well coordinated effort to remove any language barrier by having translated this guide into 22 other languages.

https://mightyohm.com/blog/2011/04/soldering-is-easy-comic-book/
http://mightyohm.com/files/soldercomic/FullSolderComic_EN.pdf

The above guide is a great way to get started especially if you are transitioning from breadboards to circuit boards. Soldering through hole parts is a specific skill which can be tricky as the goal is to heat up the leg of the part as well as the solder without overheating the part and ruining it.

Over the course of time electronic components have slowly evolved from through hole to surface mount parts. Surface mount parts can be much smaller which makes it possible for higher density boards. These parts are still soldered to the board but some of the techniques of doing so are different. A second comic book guide for soldering surface mount devices as been created by Greg Peek and Dave Roberts.

This second guide covers some of the different types of equipment needed for doing surface mount technology regardless if you are a company or a private individual. The guide also contrasts between how through hole parts compare to equivalent surface mount parts as well as how to inspect the final results.

https://mightyohm.com/blog/2012/05/smt-soldering-its-easier-than-you-think/
http://mightyohm.com/files/SMT_Soldering_Its_Easier_Than_You_Think_EN.pdf

3 SIMPLE CIRCUITS

In chapter two were a lot of parts and tools along some theory of how the parts themselves work.

The examples in this chapter start with what could be called "pure" electronic circuits which is to say just a bunch of parts with no "computer CPU like parts" to assist. These simple parts can do everything but as we transition into using the Arduino or the Raspberry Pi these parts will transition into essentially "the glue" to tie the device together to other peripherals.

Circuit with button
This circuit is a power supply, a LED, a resistor and some wire to connect it all together. This example demonstrates how the different electronic parts from the preceding chapter come together to form our simple electronic device.

This circuit is actually no different than a standard flashlight. There is a power source, a button and a LED as the light source. Pressing the button will complete the circuit and the power will flow from the battery through the resistor and back to the battery.

So in this circuit when the button is pressed the electrons flow from the negative pole of our battery through the LED then the resistor back to the positive pole of our battery.

Parts List
Item	Quantity	Description
Button	1	Push button
LED	1	5mm red LED
resistor	1	470 ohm resistor

wire	15cm	
breadboard	1	solderless breadboard
power source	1	5.5 volts

Manual Stoplight

The diagram shows our battery, a small button, and a LED. When the switch is pressed the circuit completes and the LED is lit.

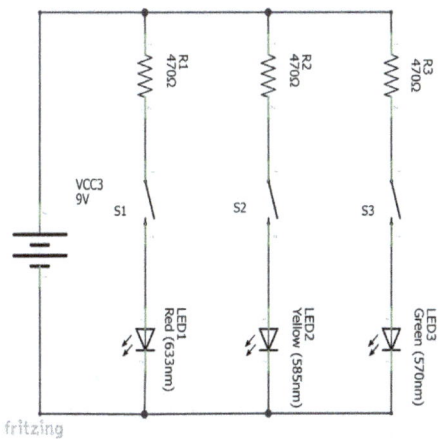

Illustration 41: Manual stoplight schematic

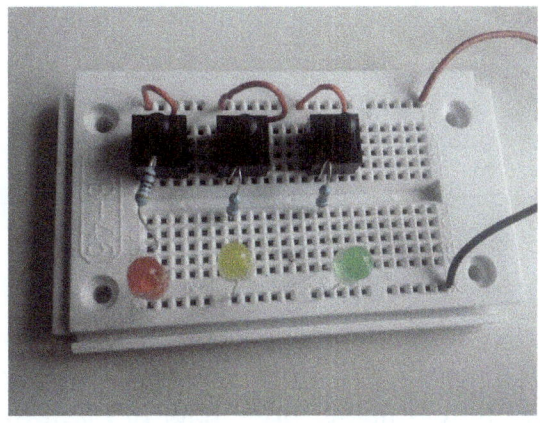

Illustration 42: Manual stoplight breadboard test

This stop light example is completely contrived as it requires manual intervention for each step. It has been included as this is actually a pretty good example that can be modified to include a micro controller to create a miniature functional stop light.

Parts List

Item	Quantity	Description
Button	3	Push button
red LED	1	5mm red LED
yellow LED	1	5mm red LED
green LED	1	5mm red LED
resistor	3	470 ohm resistor
wire	15cm	
breadboard	1	solderless breadboard
power source	1	5.5 volts

Train crossing flasher

There are probably thousands of situations where you need to create something that will work without any human intervention. An example would be some circuit that has a timing element, such as an alarm clock, egg timer or even some sort of rhythmic flashing such as a train crossing light.

This circuit will use a couple of transistors and a couple of capacitors to create a timer. This circuit only allows one LED to be lit at a time, while one capacitor is charging it will open the circuit or the opposite LED, while the capacitor is discharging it will close the circuit and allows the LED to light up. The duration will be determined by the size of the capacitors as well as the size of the resistors.

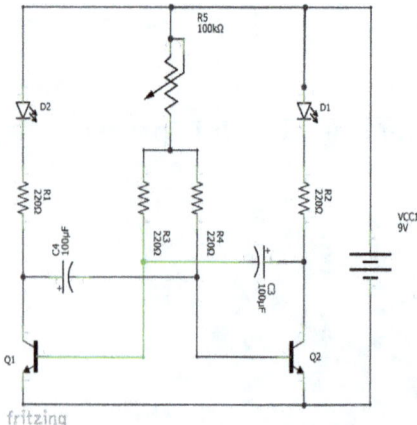

Illustration 43: Flasher circuit schematic

Changing resistors R1 or R2 will alter how quickly the capacitors will load while R3 and R4 will control how quickly the capacitors will discharge, which will affect the frequency that the LEDs will blink. It is recommended to experiment with the resistors and capacitors using different sizes to see the changes to the circuit.

Illustration 44: Flasher circuit breadboard test

Parts List

Item	Quantity	Description
red LED	2	5mm red LED
resistor	2	470 ohm resistor, R1, R2
resistor	2	4.7k ohm resistor, R3, R4
potentiometer	1	10k potentiometer, R5
BC 547B	2	NPN transistors
wire	15cm	
breadboard	1	solderless breadboard
power source	1	5.5 volts

Table 10: Parts list for train flasher circuit

The equivalent of this circuit in software would be an infinite loop. As long as the circuit is powered, the two LEDs will flash alternately forever.

The couple of circuits described thus far have been using the simplest components. No fancy integrated circuits with their fancy logic have been used but that is about to change.

Resistors uses in circuits
Resistors are used to reduce the flow of electrons to following parts of the circuit. One such use of a resistor is when using LEDs. The current is throttled in order to not burn out the LED. A resistor can also be used in conjunction with a capacitor to form a simple timer. The size of the resistor determines how quickly the capacitor will charge or discharge. This was demonstrated in the train crossing flasher example.

Using a micro-controller in your circuit actually gives you the same control over your device as a normal computer. You write programs to manipulate variables but you can also access the pins of the micro-controller. When software is reading input from an I/O port the value will be either zero or one. If that pin is not tied to either ground or to Vcc, the value is also undefined. This is the hardware equivalent of an initialized variable. This situation is referred to as floating pin. To prevent this unknown state, you can use a pull-up or a pull-down resistor to ensure that the value is defined.

Pull up resistors
Hardware, just like software, can be designed to guarantee all values read are consistent. In software, you initialize all variables prior to testing them. This same idea is done with a pull-up resistor in a circuit. The "pull-up" resistor is connected to a positive value which is the equivalent of the initialization. Illustration 45 is an incorrect implementation of this initialization. The reason is that input pins on a transistors cannot accept an endless source of power without burning out. A continuous source of a power to an input pin without any limit is a short circuit.

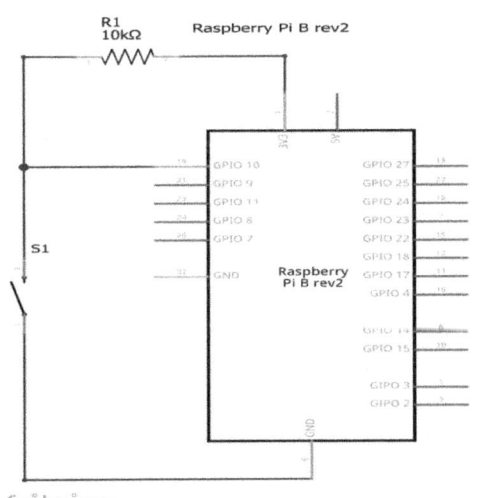

Illustration 45: Incorrect circuit with pull up resistor and button

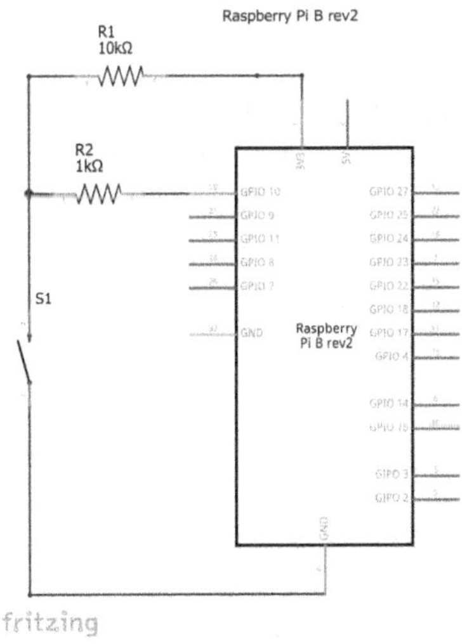

Illustration 46: Circuit with pull up resistor and button

Illustration 46 is the correct implementation for a pull-up resistor. The rest of the circuit is designed that the current will go to ground when the button is pressed. Reading from the pin when the button is not pressed will yield a positive value. However, when the user presses the button S1 it will connect the pin to ground as there is no resistance in that part of the circuit.

The program monitoring this pin will be able to determine the state of the button by the value returned when reading the pin. Pull up resistors are not only to guarantee an input value on a circuit, they are also to prevent a short circuit.

It is possible to set the value of a pin to one but it is also possible to set the value to zero through the use of a pull-down resistor. The theory is identical, yet instead of setting the value to one, it sets it to zero. Both pull-up and pull-down resistors are common but pull down are not as often used. Examples of pull up resistors can be seen in both the monostable and astable circuit diagrams described in the next chapter.

Clarification
If the GPIO pin 10 is properly programmed as input we only need resistor R1 in illustration 45, as it will be our pull up resistor and the our pin will not be floating. Yet if the developer makes a mistake and sets GPIO pin 10 to output there will be a problem. If that were to happen and the user presses the button, we will have a short circuit between GPIO 10 and ground.

By adding resistor R2 to this small circuit we guarantee no matter what the developer does pressing the button will not cause a short circuit.

Pull-up and pull-down resistors are in our circuits to ensure that it is always is a known state but also to protect the hardware from the software developers. We will be dealing a lot more with pull up resistors in chapter 7 when working with the GPIO pins and I2C programming.

Capacitors in circuits
There are a lot of general uses for capacitors from audio filtering to energy storage. There is not enough space to comprehensively go through with examples for all situations so I will pick what may be one of the most important uses – filtering unwanted noise from the power supply.

Virtually every electronic device uses direct current, and electronic components have defined ranges for the current that they use. Most of the capacitors that exist in circuits are decoupling capacitors. They will remove the small spikes or ripples that would adversely affect the proper operation or even harm the IC's in our circuit.

The decoupling capacitor is a capacitor that is connected between power and ground near either an IC or the power supply. The large electrolytic capacitor should be placed close to the power supply, its job will be to filter out low frequency noise. Yet one capacitor is almost never enough, usually a circuit will have at least one ceramic capacitor per IC in the circuit in addition to the electrolytic capacitor. The ceramic capacitor should be placed as close to the integrated circuit as possible where it will filter out the high frequency noise that is generated by the IC itself.

A second advantage to a decoupling capacitor is it will help with the providing uniform voltage to the circuit. The large capacitor will filter out the low frequency noise, but it will also temporarily provide power if the voltage drops. A decoupling capacitor is also known as a "bypass capacitor" because it bypasses the power supply when the voltage drops.

Note: When inserting capacitors you must mind how they are installed. Capacitors, with the exception of ceramic capacitors, are polarized and may explode if inserted backwards into a circuit.

Pull up resistors prevent poorly written software from damaging our hardware, decoupling capacitors ensure we have an appropriate power as input as well as ensuring we don't create any interference from our IC's, but there is more than can be done to protect our circuit.

Diodes

The most common use of a diode is to allow the current to pass in one direction only, any current that attempts to go the other way is blocked. It may seem obvious that the current will only go in a single direction and normally this is the case. Diodes can be inserted into a circuit near the power connector to ensure that if the power is connected incorrectly that it will not fry the rest of the circuit or at least not the other circuitry that is attached to it.

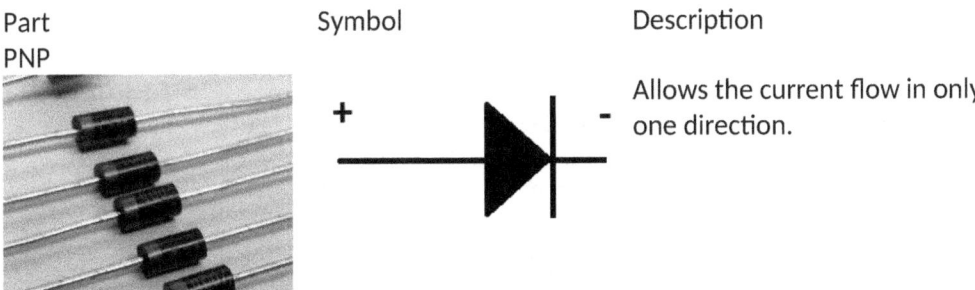

Part	Symbol	Description
PNP	+ ▶︎｜ −	Allows the current flow in only one direction.

Because of how diodes work, they can also be used in the process of rectification, which is to convert alternating current to direct current.

A simple way of remembering the orientation of the diode is to remember that the stripe on the diode matches up with the straight line on the symbol.

This is not to say that there is a part called a diode and that's it, no just like many other components there are also a number of diode variations.

Avalanche diode	These diodes will conduct in the reverse direction but only if the voltage exceeds the breakdown voltage. The advantage of this diode is that this reversing does not destroy the diode.
Laser diode	This diode which is different from the LED will convert energy into coherent light. That is all the light waves will be pointing in the same direction.
Light emitting diode	This is one of the most common diodes but mainly due to the ability to emit light in a controlled manner. These can emit light in most of the colors, infrared and also ultra violet.
Photo diode	A light activated diode.

Schottky diode	This diode is very efficient when compared to a standard diode. The voltage drop in a schottky diode may be as little as a 1/3 to 1/2 when compared to normal voltage drop of a zener diode.
Tunnel diode	These diodes are very fast and can be used in low temperatures, high radiation environments and high magnetic field. These properties make this a good choice for spacecraft.
Zener diode	The zener diode exhibits the same functionality seen in an avalanche diode, but usually the breakdown voltage for zener diodes is 5 volts while avalanche diodes have a breakdown voltage of greater than 5 volts.

Is this really necessary

This is a very dense, yet important look at why capacitors end up on circuit boards and in circuit diagrams and what role some of the resistors are playing in the circuit diagrams. It is quite possible to recreate some of these small circuits on breadboards without using some of these protective parts and they will still work, but they should not be left out when creating circuit boards.

Why is this the case? For our breadboards we will probably be using simple batteries which are providing a very steady power. It is also possible our creation is powered from a very high quality AC to DC power supply. The next power supply connected may be of a lower quality, or the next circuit may involve more sensitive chips.

The pull-up or pull-down resistors are good style, but they are also important to prevent hardware damage due to a software mistake. Software is very malleable and can take a lot of punishment, but this isn't always the case for hardware.

Rectifier

This last section on rectifiers is actually interesting and does show a direct use for diodes, as well as explain how to get direct current from alternating current. Yet beyond satisfying intellectual curiosity it will not be used in the rest of the book.

Most circuits will receive their power as direct current yet some will actually accept alternating current which may be a bit confusing considering that the electronic components require direct current.

The answer is that rather than having an external power supply or wall wart that does the conversion the device itself has the necessary components to convert the AC to DC. The neat thing is that with just the few components that we have been looking at[19] up until this point we can convert AC to DC within our own devices.

The alternating current has a wave that looks similar to Illustration 47.

---

19 Well, plus a small transformer to lower the voltage

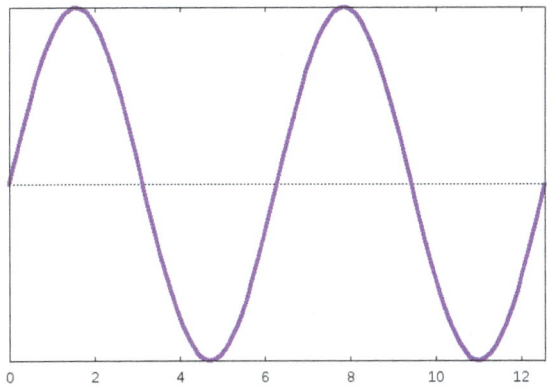

Illustration 47: Alternating current graphed over time

The first step to converting AC to DC is to eliminate the negative portion of this curve. This is actually easy enough to do using a diode. The diode will only let the positive portion of the current through.

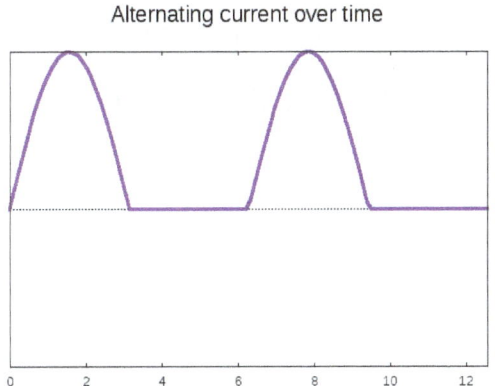

Illustration 48: Output from half wave rectifier

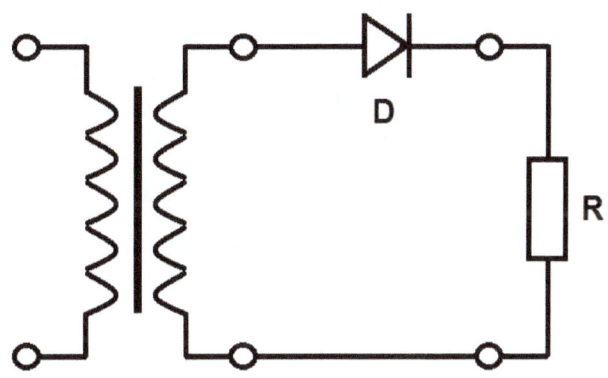

Illustration 49: Half wave rectifier

The half wave rectifier gets its name as it only half of the AC current is passed through, while the other half is blocked. The output is a direct current but it is a pulsing current. This can be corrected by smoothing or filtering the current. In general this can be smoothed out by adding a fairly large electrolytic capacitor. The capacitor will charge and discharge and in doing so it will fill in the gaps to give a fairly level supply.

A second solution for converting AC to DC is to use a full wave rectifier. As you might guess from the name, the full wave rectifier will use both the positive and negative halves of the current when creating the DC output.

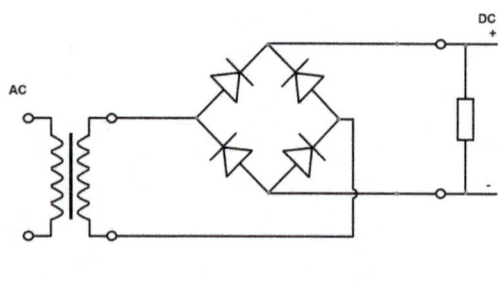
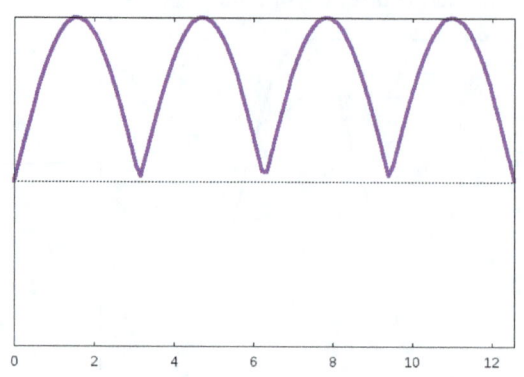

Illustration 50: Full rectifier　　　　　*Illustration 51: Output from full wave rectifier*

Again, this output is also a direct current (but pulsing current) but in order to have a nice even current this output will have to be filtered or smoothed by adding a capacitor. When a capacitor is added it will store energy during peaks cycles and release it when the signal dips. This action of the capacitor is what smooths the curve into the flat direct current that we are used to seeing.

This really does look cool but how does it work? Well, it is a very ingenious piece of work. The heart of this circuit is the diode bridge, which is when four diodes are put together so that the polarity of the output will be the same regardless of the polarity of the input. Basically the output will be positive despite the fact that the input is positive or negative.

When the input on the left of the bridge is positive and the input on the right side is negative, the positive current will flow along the red path and the negative will flow along the blue path.

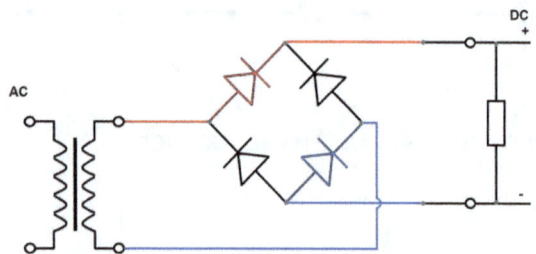
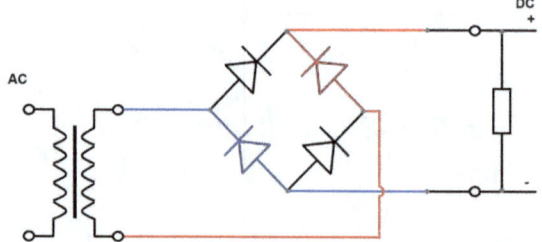

Illustration 52: Input on left side of bridge positive　　　*Illustration 53: Input on left side of bridge negative*

However, when the input on the left of the bridge is negative and positive on the right side of the bridge the positive current will flow along the red path and the negative will follow the blue path.

It is no longer necessary to build your own rectifier, it is possible to purchase a single bridge rectifier as a single part. As cool as the electronics is for converting AC to DC, in most cases you can purchase a small power supply that will produce the DC you need for just a small amount money. If you really need to keep the price down, purchase a power supply that is without a case and power wire, and built it into your project directly.

One final hint
Never underestimate the value of quality components and power supplies.

Experimenting with unfamiliar integrated circuits can be a time consuming exercise. The specification sheets are usually very long and full of various thresholds, timing or even the package definition.

It is easy to make a mistake and spend hours trying to determine where exactly you went wrong. It is also possible that a low quality part is the problem not the component that you are experimenting with.

The same is also true of your power supply or breadboard. Circuits don't behave as expected when the power is insufficient. Cheap breadboards can have poor internal connections which also can also cause hours of frustration until the problem is eventually discovered.

4 INTEGRATED CIRCUITS

The history of the computer started during wartime with their task being to calculate ballistic tables for the military. These calculating machines were big power hungry beasts that were made of vacuum tubes.

In addition to using a lot of electricity a lot of heat is also generated both then and now. It was probably the abundance of heat that caused these old machines to break down so often.

Transistors that replaced the vacuum tubes were smaller, produced less heat and were more reliable. These transistors were then connected with other discrete components. The breakthrough was to pack multiple transistors together to make a circuit in a single package.

These transistors are sometimes referred to as an integrated circuit, sometimes called an IC, computer chip or microchip is an electronic circuit. This circuit may contain a CPU, memory or may simply perform special functions. This integrated circuit is pretty fragile which is why it is encased in a plastic or ceramic package.

Some of the integrated circuits are more fun to use than others. It isn't just a few transistors or a few logic gates but actually some are complex enough that they are essentially a tiny little computer. Over the years they have become more and more complex but it is hard to overlook the contributions of some of the forerunners.

If a integrated circuit could be a classic rock star, the 555 timer IC would be it. This chip can be used in timing circuits, pulse generation and oscillator applications. It was first introduced in 1972 by Signetics and perhaps because of the low price and high reliability that makes this chip both easy to obtain and it is still popular today. If the claims are to be believed[20], perhaps 1 billion of these chips are manufactured and used each year.

---

20 http://www.semiconductormuseum.com/Transistors/LectureHall/Camenzind/Camenzind_Page2.htm

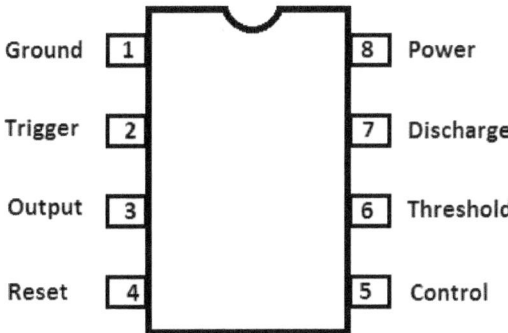

Illustration 54: 555 Timer chip

In the previous chapter a small train flasher was created using discrete components. It is possible to do much the same thing using the 555 integrated circuit. We only need to add a couple of resistors and capacitors to create a timer that can drive some other circuit, chip or in that case LEDs.

Timer chip 555

It is common in circuit diagrams to view the input and outputs of the chip differently than the actual pin layout. They are usually displayed in circuits not as a chip but rather with all of the power on top, inputs on the left side, ground on the bottom and the outputs on the right side.

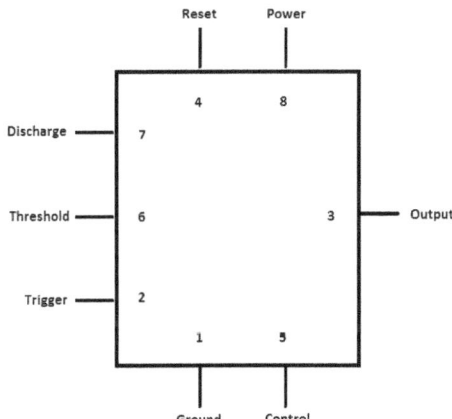

Illustration 55: 555 timer chip layout

The 555 is a very flexible chip which can be used to create several different types of timing circuits.

Monostable circuit

A simple timing circuit that you may wish to include in another project is a monostable circuit. In this circuit when we trigger the 555 it will bring the output high for the "programmed" amount of time and then the output will go low and stay low. This is basically the functionality that we see in a kitchen timer; the circuit is activated and after a given amount of time the timer is finished and an alarm will sound.

When the 555 is set up as a monostable circuit, the trigger pin will cause the circuit to enter an unstable state, which will cause the output to go high for the specific amount of time before returning to the stable state.

Illustration 56: Monostable circuit over time

In the stable state the output will go low. The time period in seconds that the circuit will be unstable is defined by the resistor and capacitor used in the circuit. A monostable circuit will have a single iteration based on the reset being triggered.

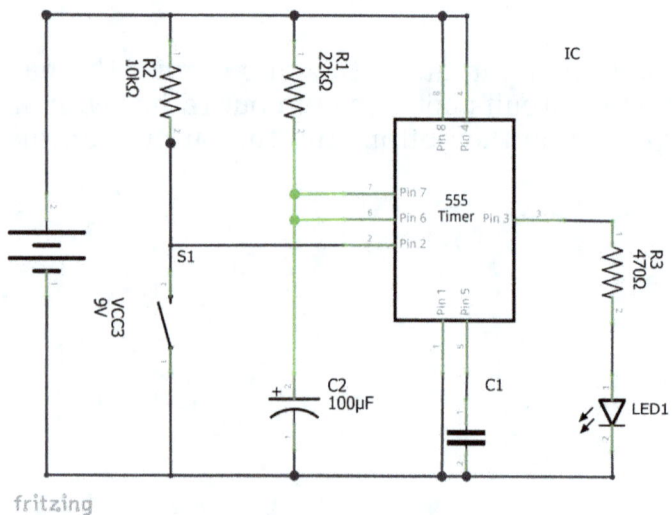

Illustration 57: Monostable circuit schematic

It is possible to precisely set the length of the cycle by selecting a specific resistor and capacitor.

$T = 1.1 \times R1 \times C1$

$R1 = 1k$ ohms

$C1 = 100uf$

$T = 1.1 \times 1000 \times 0.0001$

$T = 0.11$

Below is a small table with a few values that could be used in a monostable circuit and how long the delay would be in seconds.

R1	C1	T
1k	100uf	0.11
22k	100uf	2.42
22k	1000uf	24.2

T = time in seconds
R1 = resistance in ohms
C1 = capacitance in farads

Astable circuit

A second type of circuit that the 555 can easily support is the astable circuit. An astable circuit is a two state circuit where it is not stable in either state. The circuit will alternate between the two different states and settle down in neither of them.

Just like the monostable circuit the output of the astable circuit produces a wave but rather than a single small bump, it will produce a square wave that changes between low and high at regular intervals with the duration of these highs and lows based on the resistors selected for the circuit.

Illustration 58: Astable circuit over time

The time period in seconds for a single cycle is defined by the resistors and capacitor used in the circuit. The time for a single complete cycle, both T1 and T2, is called the duty cycle. The duty cycle does not have to be symmetric[21] although in this example it is at 50%.

---

21 Actually an asymmetric duty cycle would be an example of pulse width modulation which is another interesting topic and one that the 555 supports.

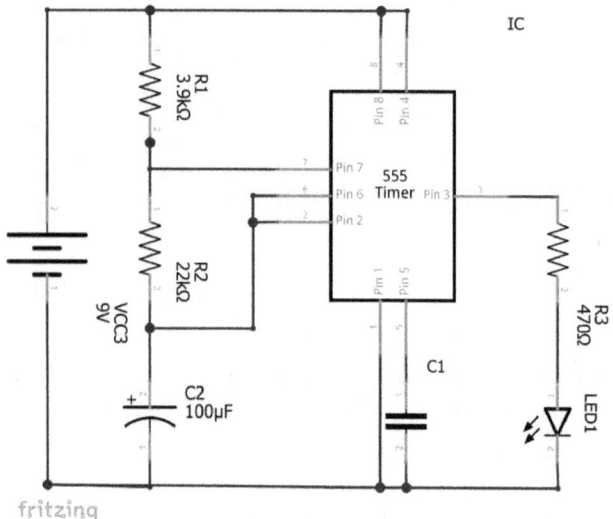

Illustration 59: Astable circuit schematic

$$T1 = 0.7 \times (R1 + R2) \times C1$$
$$T2 = 0.7 \times R2 \times C1$$
$$T = T1 + T2$$

Knowing what values of our resistors and capacitors allows us to determine how long the cycle will be.

R1 = 1k ohm
R2 = 22k ohm
C1 = 100 uf

$$T1 = 0.7 \times (1000 + 22000) \times 0.0001$$
$$T1 = 1.61$$

$$T2 = 0.7 \times 22000 \times 0.0001$$
$$T2 = 1.54$$

$$T = 1.61 + 1.54$$

Below is a small table with a few values that could be used in a astable circuit as well as how long the delays would be.

R1	R2	C1	T1	T2
1k	1k	100uf	0.14	0.07
1k	22k	100uf	1.61	1.54
3.9k	22k	100uf	1.81	1.54
22k	22k	100uf	3.08	1.54
22k	22k	1000uf	30.8	15.4

T1,T2 = time in seconds

R1   = resistance in ohms
R2   = resistance in ohms
C1   = capacitance in farads

A couple of general things to keep in mind for the astable circuit.

- a larger capacitor will increase the cycle time and thus reduce the frequency
- increasing the size of the resistor R1 will increase the time high (T1) without changing the time low (T2)
- increasing the size of the resistor R2 will increase both the time high and the time low, but will also increase the duty cycle.
- the resistor R2 needs to be much larger than R1 to get a duty cycle approaching 50%

bistable

The monostable and astable circuits are both examples of multivibrator circuit. A multivibrator circuit is a circuit that implements a two state system and both of these do that job completely but there is one more type of multivibrator circuit and that is the bistable circuit.

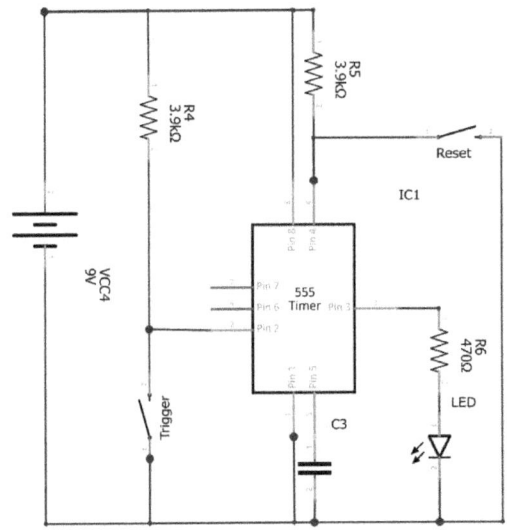

Illustration 60: Bistable circuit schematic

There is yet another type of bistable circuit. That stable also has two stable states, but the difference is that the output from pin three is fed back into the circuit. It is then possible to remove one of the switches then every time the switch is pressed the state changes.

Like the others it also implements two states but it differs in that when triggered it will switch state to the opposite state. This type of circuit could be used to store one bit of information.

The 555 integrated circuit can create a multivibrator circuit with only a few other components. Timing circuits are used in alarm clocks to microwave ovens but there is yet another integrated circuit that has both the flexibility and power that can outshine the 555

– the micro-controller.

Micro-controller

Micro-controller's are actually not too different from just a standard CPU but they are also somewhat similar to a system on a chip (SOC) - the difference is really just a matter of complexity. Both chips have a CPU, input/output ports, support for external clock crystal and memory on a single chip. Because the micro-controller isn't as complex as a SOC processor doesn't imply that it is somehow lacking in functionality.

One example of a micro-controller is the Atmel attiny85. This micro-controller has real time clocks, two wire mode bus (I2C), serial peripheral interface (SPI), universal asynchronous receiver/transmitter (UART), five input/output pins and hardware support for pulse width modulation.

This is a lot of functionality to be supported by a single eight legged chip. It is actually so much functionality that it is not possible to support it all directly with so few legs. The only way to support all of this would be to for each of the legs of the micro-controller to support multiple types of functionality. This is exactly how this has been implemented.

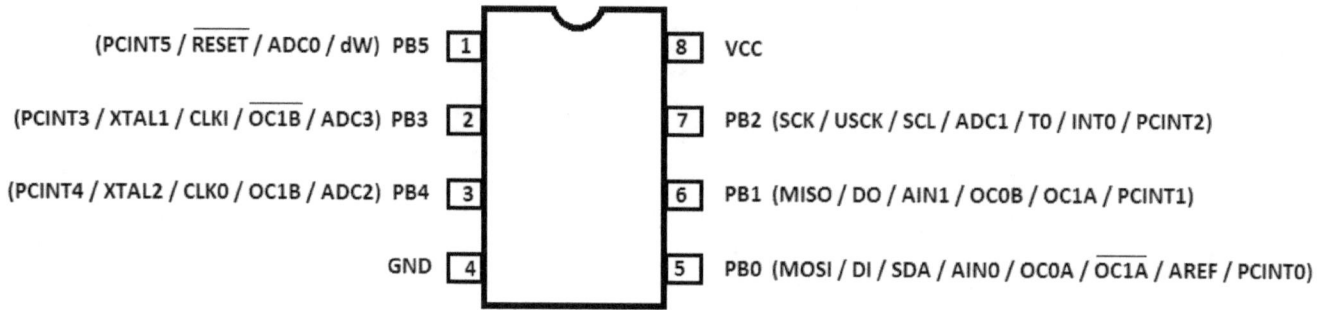

Illustration 61: Micro-controller functionality

The solution to unlocking these functions to decide which ones you need and to enable those specific ones for your project. This configuration is done by setting the fuses for the micro-controller.

Fuses, despite the name, have nothing to do with the more traditional fuses which are used to protect circuitry by melting when the current exceeds their rated level. The fuses for the Atmel micro-controller are three bytes of permanent memory that affect what functions are enabled when the chip is powered on. When we say that these "fuses" are permanent, it is different than other types of ram in the sense that they will retain their value even when the power is off.

The fuse bytes are bit mapped and can be manually calculated while reading the datasheet but a much easier method is to use a fuse calculator[22].

However, some care must be taken when deciding on which values the fuses should be set to. If set incorrectly, it may be no longer possible to communicate with the Atmel. It is possible to reset the fuses but not every programmer can reset these fuses.

---

22 http://www.engbedded.com/fusecalc

Engbedded Atmel AVR® Fuse Calculator

Device selection

Select the AVR device type you want to configure. When changing this setting, default fuse settings will automatically be applied. Presets (hexadecimal representation of the fuse settings) can be reviewed and even be set in the last form at the bottom of this page.

AVR part name: [ATtiny85 ▼] [Select] (141 parts currently listed)

Feature configuration

This allows easy configuration of your AVR device. All changes will be applied instantly.

Features

[Int. RC Osc. 8 MHz; Start-up time PWRDWN/RESET: 6 CK/14 CK + 64 ms; [CKSEL=0010 SUT=10]; default value ▼]

☐ Clock output on PORTB4; [CKOUT=0]
☒ Divide clock by 8 internally; [CKDIV8=0]

[Brown-out detection disabled; [BODLEVEL=111] ▼]

☐ Preserve EEPROM memory through the Chip Erase cycle; [EESAVE=0]
☐ Watch-dog Timer always on; [WDTON=0]
☒ Serial program downloading (SPI) enabled; [SPIEN=0]
☐ Debug Wire enable; [DWEN=0]
☐ Reset Disabled (Enable PB5 as i/o pin); [RSTDISBL=0]
☐ Self Programming enable; [SELFPRGEN=0]

Illustration 62: Fuse calculator

Manual fuse bits configuration

[Apply feature settings]

This table allows reviewing and direct editing of the AVR fuse bits. All changes will be applied instantly.

Note: ☐ means unprogrammed (1); ☒ means programmed (0).

Bit	Low	High	Extended
7	☒ **CKDIV8** Divide clock by 8	☐ **RSTDISBL** External Reset disable	
6	☐ **CKOUT** Clock Output Enable	☐ **DWEN** DebugWIRE Enable	
5	☐ **SUT1** Select start-up time	☒ **SPIEN** Enable Serial Program and Data Downloading	
4	☒ **SUT0** Select start-up time	☐ **WDTON** Watchdog Timer always on	
3	☒ **CKSEL3** Select Clock source	☐ **EESAVE** EEPROM memory is preserved through the Chip Erase	
2	☒ **CKSEL2** Select Clock source	☐ **BODLEVEL2** Brown-out Detector trigger level	
1	☐ **CKSEL1** Select Clock source	☐ **BODLEVEL1** Brown-out Detector trigger level	
0	☒ **CKSEL0** Select Clock source	☐ **BODLEVEL0** Brown-out Detector trigger level	☐ **SELFPRGEN** Self-Programming Enable

Illustration 63: Fuse calculator continued

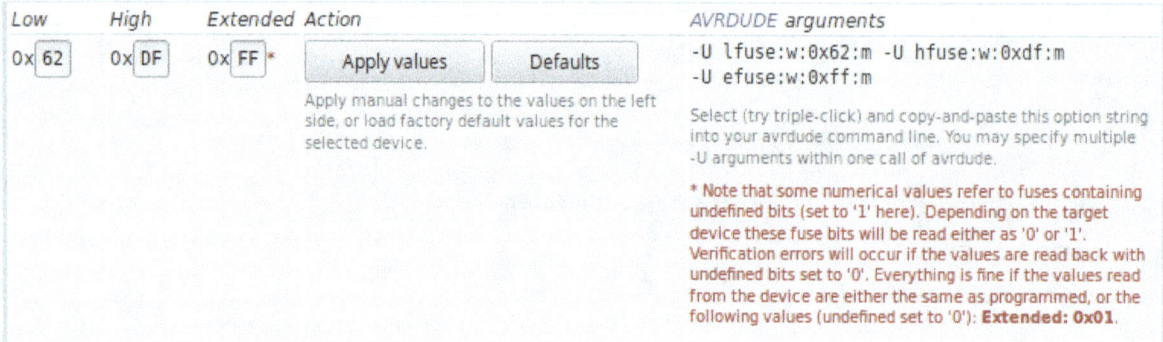

Illustration 64: Fuse calculator continued

Note: Do not modify your own fuses unless you are certain you know what you are doing. If you pass in incorrect values it may no longer be possible to communicate with your micro-controller.

This may or may not be correctable with your micro-controller programmer, but, your mileage may vary. The more expensive micro-controller programmers can reset fuses.

Memory

The Atmel is from a family of 8 bit RISC micro-controller developed in 1996. One of the things that made this family of micro-controllers unique, well at that time, was its use on-chip flash memory for program storage. The other micro-controllers at that time were using one time programmable ROM, EPROM or EEPROM memory.

The Atmel AVR family of micro-controllers have three different types of memory.

- Program
- Data
- EEPROM

Program memory

This is a fairly large piece of memory (e.g 8kb) which is used to hold the program for the micro-controller. This is usually referred to as the flash memory. This memory is only writable a reasonably small number of times.

In the example of the Atmel attiny85, the number of write/erase cycles to the flash memory is limited to about 10,000. Although this number may sound small this should be more than enough times during development to get a working product and once you move to production this number will not come into play.

Data memory
Data memory contains data such as pre-initialized variables anonymous strings or other temporary data. Depending on how the program was coded other constant data may be located in either the program memory or in the data memory. This is a useful trick to maximize the memory usage.

In addition to this type of data, the data area also contains registers, input/output space and special function registers. The input/output space is used by the chip when doing SPI or I2C data transfers.

EEPROM
The EEPROM is yet another separate data area where you can store data or configuration for your specific device. You can access this memory directly reading or writing single or multiple bytes. One unique limitation is that this memory also has a limited number of write/erase cycles which for the attiny85 is about 100,000. This number is fairly low if you are writing to your EEPROM every time through a loop, but this should be more than enough write cycles to store basic configuration setup. It could even be modified once a day and still provide a lot of error free use for the next two to three hundred years.

Exception to the rule
One of the important differences between a device with a micro-controller versus a computer with a cpu is the operating system. This is in part because micro-controller have a tendency to be both slower and to have less resources (e.g ram) available to hold an operating system.

Yet the history of computing started with much more modest resources than what is available today. Some of the micro-controllers today have more resources than the early computers. Those computers did not have the amazing GUI's but were quite capable in their own right. The 16 bit Apple operating system called ProDOS which powered the Apple IIgs required only 64kb. That operating system supported the access between the computer and the disk (or hard disk) but also had some rudimentary operating system features.

Despite having the words operating system in the name ProDOS would be pretty weak when compared any of the current operating systems. This operating system was mainly for interacting with the disk and quite a bit less as a proper operating system.

A proper small real time operating system does exist – it is called FreeRTOS[23]. This operating system has been developed over the last 15 years in partnership with leading chip companies. This OS can be used on micro-controllers and small microprocessors. FreeRTOS compares very well with any of the historical operating systems such as MS DOS or ProDOS.

Features
  • Pre-emptive or co-operative operation
  • Very flexible task priority assignment

[23] https://www.freertos.org/Documentation/161204_Mastering_the_FreeRTOS_Real_Time_Kernel-A_Hands-On_Tutorial_Guide.pdf

- Flexible, fast and light weight task notification mechanism
- Queues
- Binary semaphores
- Counting semaphores
- Mutexes
- Recursive Mutexes
- Software timers
- Event groups
- Tick hook functions
- Idle hook functions
- Stack overflow checking
- Trace recording

Controlling many pins

It would not be very efficient to have a very small micro-controller with hundreds of pins. It is possible to have one with 50 but this micro-controller would have a very different profile. It would have more memory, timers, I/O ports and would cost more. It would also on average be about the size of a small to medium CPU.

This problem was experienced and solved long before micro-controllers were in every circuit and device. This was solved by having other chips which would control more pins while just using a few.

7 Segment driver

A rather fun example of this is the 7 segment driver chip. This chip was designed to convert between a binary coded decimal input and another format. In this case it was intended to provide the interface between the circuit board and a 7 segment display.

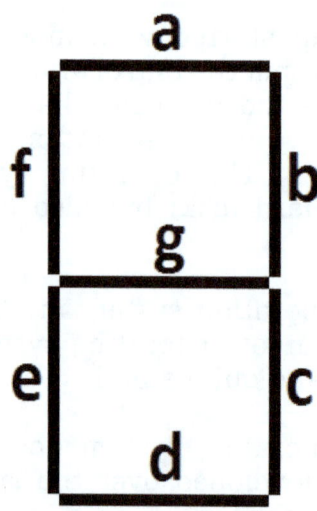

Anyone who has a microwave or a DVD player should be familiar with this particular output. The 7 segment display is just a very fancy package of seven LED's with each one being addressable.

This is terribly flexible but perhaps less convenient for the developer who doesn't want to think about which pins should be lit up on this chip but rather simply wants to display the value "5".

Individual Segments							Value Displayed
a	b	c	d	e	f	g	
*	*	*	*	*	*		0
	*	*					1
*	*		*	*		*	2
*	*	*	*			*	3
	*	*			*	*	4
*		*	*		*	*	5
*		*	*	*	*	*	6
*	*	*					7
*	*	*	*	*	*	*	8
*	*	*			*	*	9
*	*	*		*	*	*	a
		*	*	*	*	*	b
*			*	*			c
	*	*	*	*		*	d
*			*	*	*	*	e
*				*	*	*	f

Table 11: *Segment setup for 7 segment display*

The 74hc47 chip comes to the rescue. This chip will take as input a four bit BCD coded value and will use that to send the power to the correct segments of the 7 segment display for each value.

Decimal	Binary Pattern				BCD
0	0	0	0	0	0
1	0	0	0	1	1
2	0	0	1	0	2
3	0	0	1	1	3
4	0	1	0	0	4
5	0	1	0	1	5
6	0	1	1	0	6
7	0	1	1	1	7
8	1	0	0	0	8
9	1	0	0	1	9

Table 12: *Binary coded decimal values*

Unlike the seven segment display, the BCD format only supports the decimal numbers from zero through nine.

3 to 8 decoder / demultiplexer

There are probably many other types of supporting IC's but another interesting one is the 74HC138 3 to 8 decoder. This chip acts somewhat similar to a I/O expander. It is possible to control eight output pins with only three input pins.

Yet this is slightly misleading. It is possible to decode a three bit value so one of eight output pins will be enabled. This is pretty neat but it is somewhat limiting. This type of chip might be necessary (along with a multiplexer) in a circuit that is transferring data between two points.

4 to 16 decoder / demultiplexer

This chip, 74HC4514, is actually quite similar to the 3 to 8 decoder chip, it is just on a slightly larger scale. It is possible to create a 4 to 16 decoder using either two 3 to 8 decoder chips.

8 bit shift register

The 3 to 8 decoder is a convenient way to use a small number of pins to control eight different pins. The downside of the decoder chip is that it only enables one of eight pins based on the three input pins. In some situations this might be good enough but if you need to simultaneously turn on more than one pin.

This is possible with a completely different integrated circuit – the 8 bit shift register. This chip is perhaps a small bit less comfortable to use. Not because of how many pins are required, requires one pin for input instead of three but it takes the input one bit at a time.

The 74hc585 shift register all eight bits need to be passed in separately. This means the chip is essentially a serial to parallel converter. The good news is that this makes it possible to potentially set all eight pins active. If this chip is being used to light up LEDs then this chip would allow us to light them showing the binary value of the byte being processed.

```
int dataPin = 4;
int latchPin = 5;
int clockPin = 6;

byte ledpattern1 = 0xC3;
byte ledpattern2 = 0x55;

void setup()
{
  // set three of our micro-controller pins to output
  pinMode(latchPin, OUTPUT);
  pinMode(dataPin, OUTPUT);
  pinMode(clockPin, OUTPUT);
}

void loop()
{
  // switch between two different patterns
  loadShiftRegister(ledpattern1);
  delay(500);
  loadShiftRegister(ledpattern2);
  delay(500);
}

void loadShiftRegister(byte val)
{
  // pull latch low
  digitalWrite(latchPin, LOW);
  // load each of the bits into the chip
  shiftOut(dataPin, clockPin, LSBFIRST, val);
  // bring latch high causing output from chip
  digitalWrite(latchPin, HIGH);
}
```

16 bit I/O Expander

The chips that have been discussed so far are a convenient way to extend the output from a few pins to many. This is actually a strength but all of these examples are for extending the output reach of a micro-controller. It is possible that many solutions require only output but it is a bit naive to think that this is the only situation.

The Atmel attiny85 is a powerful little processor – the specifications show the attiny85 chip, Illustration 61, has a similar level of power as the Apple II computer, Table 1, when it was first released. Despite this rather amazing comparison the Atmel attiny85 does has one short coming. It is limited to how many devices it can control due to its limited number of pins. As obvious as it sounds, the solution to this general problem was the creation of a chip that can simply be used to extend the number of pins that can be controlled.

One such I/O expander is the MCP23017 from the Microchip Technology Inc. This I/O expander is not just a dumb piece of silicon either. It also supports the I2C bus protocol for both setup and control of this chip.

Illustration 65: I/O Expander

Despite how common this sounds, it is actually very powerful to have a chip or circuit control many different data lines. If controlling 16 is good then controlling more must be better. This possibility has also been considered through the I2C protocol. The chip itself has three pins which can be configured to be part of the address for the I2C protocol.

These three pins, counting in binary, allow for up to eight 23017 chips to be on the same circuit at the same time. This is done by tying these three address pins to either ground or power (Vdd) in various combinations to create an address from these three bits.

This ability for the engineer to define the address allows the him or her the ability to control up to eight different 23017's which in turn allows control of up to 128 different pins for input or output. As the name implies, this chip is for input and output. The pins from this chip can be used as either two banks of eight pins but can also be treated as a single bank of 16 pins.

Not all companies use the I2C protocol but the good news is that Microchip Technology actually
produces two different versions of this chip only differing with the supported bus protocol.

    MCP23017   16 bit I/O expander with serial interface (I2C)
    MCP23S17   16 bit I/O expander with serial interface (SPI)

This allows the engineer to pick which protocol fits best with their board. Both of these protocols will be fully described later.

Controlling the mcp23017

The advantage of the I2C protocol is that there can be multiple different devices connected together with each one listening for their own commands. The disadvantage is that there is a bit more overhead when trying to communicate with a single device.

Writing a single byte data to a device requires both the address of the chip as well as the destination register. This requires that three bytes are sent down the wire to the device, the first is the address of the device including operation type, the second is the destination register and the third is the value being sent.

The variable part of the device address is defined on the chip by the A0,A1 and A2 which will have been tied to either power (Vdd) or ground (Vss). The complete address is defined with the first four bits being 0100[24], followed by the three address bits finally with the operation bit.

Device Address

0	1	0	0	A2	A1	A0	R/W

Write = 0, Read = 1

This I/O extender is a complex device which has 22 registers for controlling the semiconductor. It is possible to setup the pins for input or output either as two eight bit registers, PORTA and PORTB, or one sixteen bit register.

It is possible to configure I/O direction, pull-up resistors, interrupts, and default register values just to mention a few. Below is a small table with registers names, address and the Power On Reset (POR) values.

Register Name	Addr (hex)	POR/RST Write = 0, Read = 1
IODIRA	00	11111111
IODIRB	01	11111111
IPOLA	02	00000000
IPOLB	03	00000000
GPINTENA	04	00000000
GPINTENB	05	00000000
DEFVALA	06	00000000
DEFVALB	07	00000000
INTCONA	08	00000000
INTCONB	09	00000000
OCON	0A	00000000
OCON	0B	00000000
GPPUA	0C	00000000

[24] For this particular I2C device the address bits is 0100 but other I2C integrated circuits will use different values.

GPPUB	0D	00000000
INTFA	0E	00000000
INTFB	0F	00000000
INTCAPA	10	00000000
INTCAPB	11	00000000
GPIOA	12	00000000
GPIOB	13	00000000
OLATA	14	00000000
OLATB	15	00000000

Table 13: IOCON.BANK = 0

This table only displays the register values required when BANK is set to zero, which is the default when the chip is first powered on. When the BANK equals one, the registers are separated into two sets of eight bit registers.

Reading a single byte data from a device requires both the address of the device as well as the source register. This requires that three bytes are sent down the wire to the device, the first is the address of the device including operation type, the second is the source register and the third is the source register again with the R/W bit set for reading.

It is not just a good idea to read the data sheet for the MCP23017 but a requirement for anyone wanting to more than just a few basic operations with this device. The sheet explains the different modes and configuration options in much greater detail that can be covered in this section.

The nice thing about this design using the I/O expander is that the entire design is very simple. We simply need to provide ground, power, and the two data lines for I2C (SCL and SDA) from the chip or other device.

Projects
All of these parts allow you to create some interesting projects. Below is a contrasting example of using some traditional integrated circuits versus using a micro-controller.

555 in action
As previously mentioned, the 555 microchip is a very versatile component to have in your tool box. Despite its apparent simplicity it can be used in circuits designed to solve any of the following tasks.

    Pulse width modulation
    Multivibrator circuits
    Emergency flashers
    Signal generator
    Police siren
    Frequency divider circuit
    Computer voice changer circuit
    Coin tosser
    Random number generator

Wiper delay

Although the 555 can be used for solving any number of serious problems, this next example will be to implement a simple flasher to make a child's toy ambulance more real by adding flashing lights to it.

Illustration 66: Modified toy ambulance

Of the three multivibrator circuits that were previously described, it sounds like a astable circuit is the one that would best fit this problem. We can use the output of the astable circuit to flash either one or two LEDs.

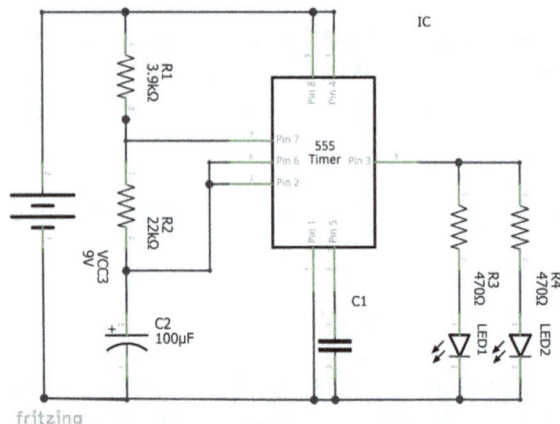

Illustration 67: A stable circuit using 555 timer

This circuit is perhaps a slightly different from the astable circuit described earlier but not dramatically so. The only difference between the two is that the output from pin three will be going through two LEDs.

If we implement this circuit we two get a simultaneous flashing LEDs where we can control the flashing frequency by varying which resistors are used. Yet it is possible with a very slight change to this circuit we can get the two LEDs to flash alternately. It is purely a matter of taste but I think that would be the more interesting solution.

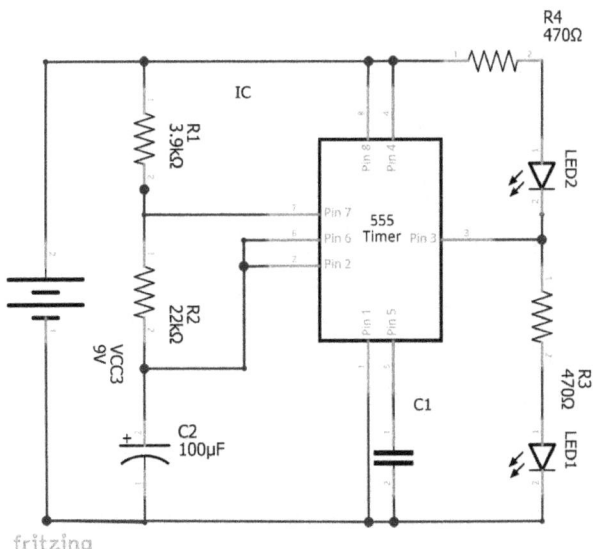

Illustration 68: Modified astable circuit schematic with alternate flashing

In a astable circuit the 555 outputs a square wave, depending on which resistors used depends on what the duty cycle will be. The LED labeled LED1 will blink just as it always has in a astable circuit. The clever bit is how LED2 is added.

The regular VCC will power LED2, however, the current cannot flow through LED1 when the output of our 555 pin 3 is high, when this is the case the current will flow through LED1 and LED2 will remain unlit. When the output from pin 3 is low, the current will flow from VCC through LED2 and back through the 555, and no current will flow through LED1. Thus this construction will give us the alternating flash that is similar to a real ambulance.

It is possible that this particular child's toy was originally envisioned to support some extra electronics. The lights on the top of the car are already hollow and have enough space for one 3mm LED on each side, and the headlights of the car are transparent from the inside. It would be possible to add a couple of white LEDs to provide light here too.

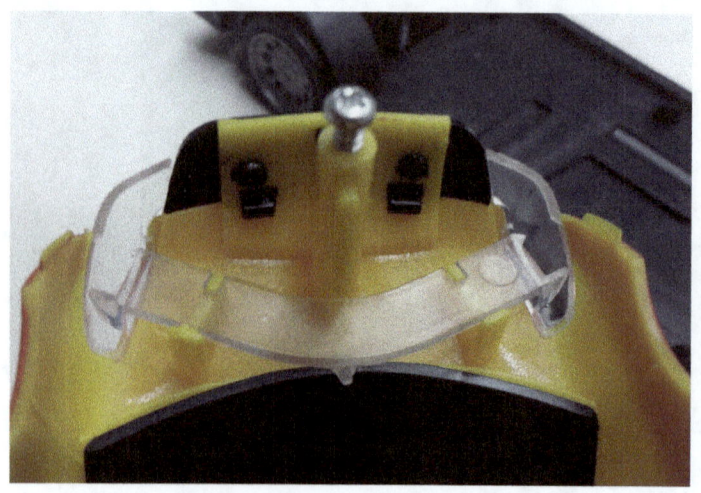

Illustration 69: Ambulance front end assembly

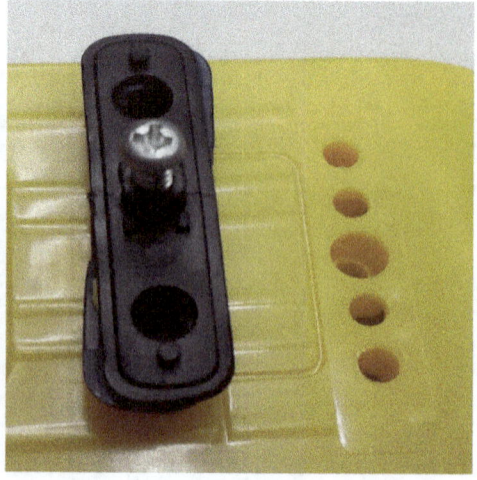

Illustration 70: Location for flashing lights

Despite the best laid plans, things do not always work out as intended. This is a simple circuit but even if everything is correctly connected up the results may not always work out as planned.

Illustration 71: Flashing circuit

While doing this project I encountered a few minor problems. None were problems with the circuit but were actually small issues that can occur when going from schematic to board. I have listed the problems and their solutions below.

Problem
The LED connected to pin 3 is always on

The LED connected to pin 3 is always on
The LED never flashes

The LED never flashes.

Solution
Check the values of your resistors and capacitor. If the values are too small the LED will be blinking so quickly it may appear to be constantly on.
The capacitor is either improperly connected or is bad. It is possible there is a loose connection with the power wire.
Is the 555 heating up? It is possible the LED is not flashing as there is a short circuit. Review your circuit against your diagram, and check each of your solder joints for either a failed connection or an inadvertent connection.

It goes without saying that it is a good idea, when possible, to test out your circuit using a breadboard first. It takes a minute and in addition to proving the circuit it also is a quick sanity check of the components. If you don't take the components from the breadboard and use them directly on your circuit board, it may be a good idea to quickly test out each component before putting it into your creation.

Item	Qty	Description	Est.Cost
IC	1	555 timer chip	0.60
LED	2	3mm blue LED's	0.40
Resistor	1	3.9k ohm resistor	0.20
	1	22k ohm resistor	0.20
	2	470k ohm resistor	0.20
Wire	40cm	0.60 mm solid wire	0.20
Capacitor	1	100uF electrolytic	0.75
	1	100nF ceramic	0.05
PCB Board	1	40mm x 70mm	0.60
Battery connector	1	9V battery clip connector	0.50
PCB Board	1	40mm x 70mm	0.60
Total			3.70

Table 14: Parts list for 555 flasher

The costs are estimated as normally you cannot purchase some of these components individually from some online sales outlets. Shipping costs can dwarf the costs of the components themselves.

Micro-controller driven flasher
As a software developer it is possible to create the exact same effect using a micro-controller. This is because the micro-controller is essentially a very tiny cpu. With this in mind it would be a very simple program to turn the LED's on and off.

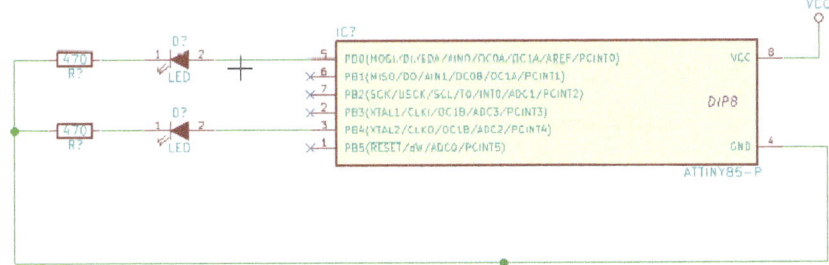

Illustration 72: Micro-controller driver flasher circuit schematic

The circuit diagram shows that all you need to do is connect the LED's to some of the micro-controller legs and add power.

From looking at the breadboard test of the circuit it should be fairly obvious that having virtually the same number and type of components that this board is actually easier to wire up than the one using the 555 chip.

Illustration 73: Breadboard implementation of micro-controller driven circuit

However, one of the "small" differences between the two circuits is that the second one actually needs a small program in order to function. The entire program listed below is quite easy to read and understand.

Code	Explanation
| ```c
#ifndef F_CPU
#define F_CPU 1000000UL
#endif

#include <avr/io.h>
#include <util/delay.h>

void initialize()
{
  // set port for writing
  // initialize port B
  DDRB =
    1 << PB0 |   // PB0 = High
    1 << PB4;    // PB4 = High
}

void blink_LEDs()
{
  while (1)
  {
    PORTB = 1 << PB4; // PB4 = High, PB0 low
    _delay_ms(500);   // sleep 500 ms
    PORTB = 1 << PB0; // PB0 = High, PB4 low
    _delay_ms(500);   // sleep 500 ms
  }
}

int main(void)
{
  initialize();
  blink_LEDs();
}
``` | // or a different frequency<br>// depending on your part<br>// for _delay_ms()<br><br>Comments for future developers<br><br><br><br><br><br><br><br>Setup the Pins 5<br>and 3 for output.<br><br><br><br>Do following loop forever<br><br>turn on pin 3 (PB4)<br>wait 500 milliseconds<br>turn on pin 5 (PB0)<br>wait 500 milliseconds<br><br><br><br>Program starts here,<br><br>calls the initialization routine<br>calls routine to blink LEDS |

The simplicity of the micro-controller circuit has the trade-off of additional complexity in the transfer of the program from the computer to the micro-controller. There are a number of different ways to compile and program micro-controllers. You can use an IDE which is an editor, compiler and transfer program all in one or you choose to do each step
```

separately from the command line.

A second difference between the two hardware solutions is the additional complexity necessary to program the micro-controller from the computer.

It may sound a bit counter intuitive that running the programs from the command line is the easier solution. Once you have connected up the tested your in-circuit serial programmer, Illustration 80 and everything works then perhaps an IDE is more convenient. However, when first setting up your tool chain it is easier to see and diagnose problems directly without them being translated or misinterpreted by the IDE.

Depending on your setup you may be able to simply attach a flat cable but it is also possible to simply connect a couple of wires to the programmer and connect them directly to the chip via a breadboard.

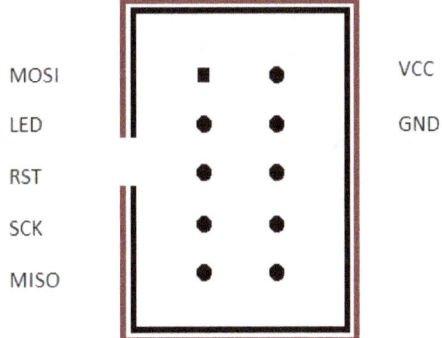

Illustration 74: 10 pin connector layout of ISP connector

One bit of good news is that the command line does lend itself well to being integrated by the make command.

Makefile for micro-controller flasher

```
DEVICE     = attiny85
CLOCK      = 1000000
PROGRAMMER = usbtiny
FILENAME   = ambulance
COMPILE    = avr-gcc -Wall -Os -DF_CPU=$(CLOCK) -mmcu=$(DEVICE)

all: clean build upload

build:
        $(COMPILE) -c $(FILENAME).c -o $(FILENAME).o
        $(COMPILE) -o $(FILENAME).elf $(FILENAME).o
        avr-objcopy -j .text -j .data -O ihex $(FILENAME).elf $(FILENAME).hex
        avr-size --format=avr --mcu=$(DEVICE) $(FILENAME).elf

upload:
        avrdude -p t85 -c $(PROGRAMMER) -U flash:w:$(FILENAME).hex:i

clean:
        rm ambulance.o
        rm ambulance.elf
        rm ambulance.hex
```

Item	Qty	Description	Est. Cost
IC	1	ATTiny85	3.00
LED	2	3mm blue LED's	0.40
Resistor	2	470k ohm resistors	0.20
Wire	40cm	0.60 mm solid wire	0.20
PCB Board	1	40mm x 70mm	0.60
Battery connector	1	9V battery clip connector	0.50
Total	4.9	Total	4.9

Table 15: Parts list for micro-controller driven flasher

The costs are estimated as normally you cannot purchase some of these components individually from some online sales outlets. Shipping costs can dwarf the costs of the components themselves.

Quite a few parts are identical between the two circuit solutions but the most visible difference is the cost of the micro-controller which actually costs almost as much as the entire 555 circuit design from table 14.

Don't overlook that a micro-controller driven solution does have one other advantage over a pure simple component solution. It is possible to update or reprogram the micro-controller if necessary. This could be to fix a bug or add new functionality. If something similar were necessary for the 555 flasher solution it might be necessary either for a lot of PCB rework or potentially replacing the entire PCB.

Always exceptions

Working with micro-controllers is actually quite nice. The circuits seem to be much simplified when compared to circuits composed of only simple through hole parts. When using a socket for the micro-controller it is even possible to do a "brain" transplant in minutes with very limited tools. Not only that, it is possible to add a header to the circuit board so the micro-controller can be programmed in place.

Having said all of that we should not forget all of the technology that made its way into our homes before these miracle devices were invented. If we just take a second look at the train flasher circuit in chapter 4, illustration 43, we will discover it is possible with a few small substitutions to come up with a solution using old discrete components with an even smaller bill of materials.

Item	Qty	Description	Est. Cost
LED	2	5mm red LED	0.40
resistor	2	470 ohm resistors, R1,R2	0.20
resistor	2	4.7k ohm resistors,R3,R4	0.20
Potentiometer resistor	1	10k ohm resistor F5	0.20
BC 547B	2	NPN transistors	0.10
Wire	15cm		0.10
Breadboard	1	Project board	0.60

Project board
Power connector   1   5.5 volts   0.50
Total   2.3   Total   2.3

Table 16: *Modified parts for train flasher circuit*

5 SOFTWARE TOOLS

There is no "right" platform to be a maker of the next "amazing device" nor is there the best development platform. Statistically speaking Microsoft windows is the most installed operating system so in general I will be describing setup and tool usage for the Windows operating system but I will also be covering Linux.

This is not because of the vast install base of Linux. Linux is probably only installed on perhaps 3% of the home personal computers. However, Linux itself is a good maker platform because of the wide variety of tools and applications that are freely available.

Linux operating system
Although people refer to Linux without much additional clarification it isn't the single homogeneous operating system that the name implies. A Linux distribution is the combination of the Linux kernel, package management system as well as the software that is also bundled with it. Although anyone can create their own Linux distribution there are quite a few popular distributions in existence so it possible to easily find one that you are comfortable with no matter what your level of computer expertise.

Over the years Linux has gotten much easier to install for both technical and regular people but it can still be a bit tricky. For a number of years the king of the hill for ease of install Linux has been Ubuntu but over the last few years Linux Mint has over taken Ubuntu in this respect.

Quite a few Linux distributions have live a boot option so Linux can be run without installing it to the personal computer. In addition to this choice it is also possible to install it in a dual boot option so either Windows or Linux can be started at boot time. The dual boot option is usually the more desirable choice for new users as they don't have to leave behind all the familiar windows all at once.

Installing Linux is outside the scope of this book but it is both free and flexible working platform with all the tools used in this book as well as a myriad of developer tools. There is an entire eco system supporting all aspects of Linux from development to community assisting users with problems.

Linux distributions

    Mint   https://linuxmint.com/download.php
    Centos   https://www.centos.org/download/

| OpenSUSE | https://download.opensuse.org/ |
| Debian | https://www.debian.org/ |

The internet is full of information on how to install Linux. Some of this may be in the form of videos with others being either the distribution sites themselves or other fan sites.

Computer Aided Design

Computers and electronics were originally designed with essentially paper, pencil, T-square, Triangle, a drafting table and an eraser. These tools in the hands of skilled designer allowed both electronic devices and printed circuit boards to be developed.

The process was obviously manual and labor intensive. The beginnings of computer aided design on a computer was just getting started in the early 1960's. Due to the cost of computers the main adopters were large aerospace companies as well as large automotive companies.

These early CAD tools did help in the design but due to the cost and the high learning curve it is likely that computers didn't speed up the original design of new products, however probably did help considerably for subsequent revisions.

Over the decades computing costs have fallen through the floor. A very nice personal computer today is used by one person while a computer from the 60's would be more than an order of magnitude slower and would be shared by many people.

Today we have both the computing power and software tools which can harness that power. Today's computer design tools are integrated affairs that manage libraries of parts, allow the user to design the schematic and then design the printed circuit board layout using the schematic as input.

Schematic editor

It is not necessary to have a special program to create the schematic for small electronic projects but as the complexity increases this type of program gives a lot of support to the engineer. It is not only about drawing lines on the screen.

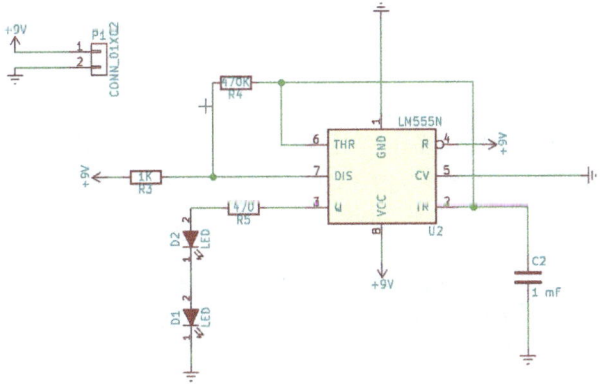

Illustration 75: Sample schematic in schematic editor

The schematic is not simply a bunch of lines, it is the map for the electrical connections between components. The schematic is actually a "map" that shows the relationship. It is possible for the software to perform checks to verify that the components are correctly connected up based on their electrical description.

The schematic itself describes the relationship between the parts in the circuit but has no connection to what shape or size a component is. Another important feature of modern CAD software is that you can do the layout of the PCB for the schematic.

PCB Layout
The schematic from above has been developed into this PCB layout using the component footprints from the library of parts. These footprints are the physical representation of the part so it is possible to do a very accurate board layout.

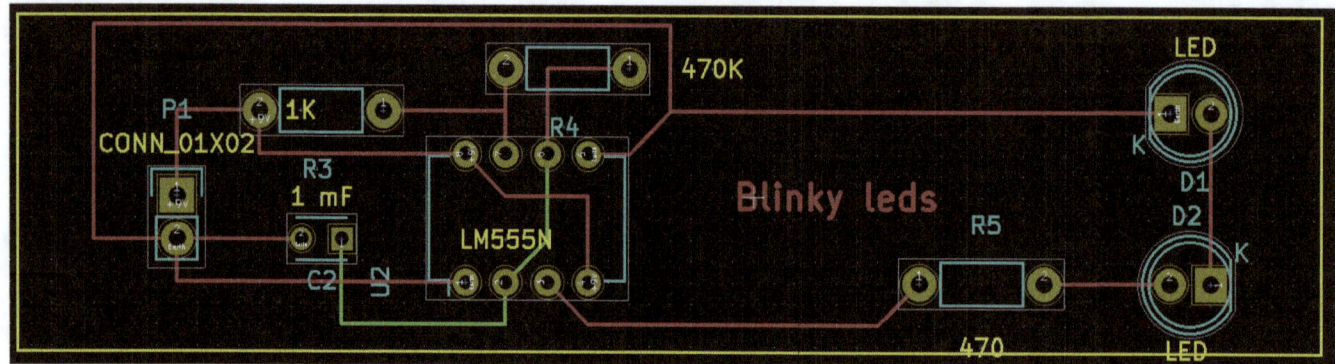
Illustration 76: PCB layout of an astable 555 circuit

Each component from the schematic has a number and thus it is possible to place each item on the PCB board while maintaining the relationship with the original schematic.

Just like your favorite word processing suite there are a few different options for procuring your favorite programs – either open source or proprietary vendor. Simply searching the internet for "electronic design automation" will return a list of at least 10 programs any of which could be used to create a basic PCB. I haven't tried all of the programs but I have tested out a few of them.

Eagle
The design tool Eagle was acquired by Autodesk in 2016. Their tool has a few different versions with the premium version supporting the development of boards of an unlimited size with up to 16 different layers. They also provide a free version of the tool which is a bit more limited in scope.

This free version, Eagle free, supports two layers and board that are no larger than 80 square centimeters in size. The other difference between the free and the non-free versions is that the non-free version has a yearly subscription fee.

This was the tool that I first started to use when trying to create my own PCB's. My mentor was able to create great work with the free version but this tool may not be the best version for the amateur just getting started. Perhaps this has changed in the intervening years both due to the inevitable enhancements to the product but also due to the increasing amount of information available on the Internet.

Operating systems
Windows
Linux
MacOS

https://www.autodesk.com/products/eagle/free-download

Fritzing
Another open source design tool is Fritzing which was developed at the University of Applied Sciences of Potsdam. The goal of the Fritzing project being to help designers move beyond experimenting with only a prototype and allowing them to create a more permanent circuit.

The Fritzing tool does have the unique feature of displaying a breadboard that is similar to how both Kicad and Eagle support creating a PCB from a schematic. It is possible with Fritzing to create your schematic but in addition to creating a PCB it is possible to also create a breadboard view using the same components keeping in mind the schematic.

Illustration 77: Breadboard view of circuit schematic in fritzing editor

Operating systems
Windows
Linux
MacOS

http://fritzing.org/download/

Kicad
The open source world also has created a few candidates which also can be used to create a PCB – Kicad. Kicad can also create boards that are 16 layers that are 2x2 meters in size. This may not be exactly the same as the premium version of Eagle but it does show that Kicad is a serious contender in this space. This software is actually so professional that it can be used by organizations with a serious agenda. One of those organizations is CERN which is home to the worlds largest particle accelerator.

Operating systems
Windows,
Linux
MacOS
http://kicad-pcb.org/download/

Programming a micro-controller

The number and types of computer solutions is a wide spectrum that covers a lot of different preferences. For computers this usually means from command line up to a GUI where nothing more than typing and clicking is the solution.

I find the command line to be a nice solution as it is usually easy to see exactly where something fails. I will be describing both the command line solutions as well as the all-in-one solutions.

Command line Compiler

When studying software development in school we spoke about a lot of things but when we spoke about the compiler it meant only one thing. It was the program that converted source code on the machine I was logged into into machine code for that same machine.

That is indeed the definition of what a compiler does but there are times that this is not possible. This could be because the destination platform is not powerful enough to run its own compiler or perhaps that target computer is a brand new chip. The solution is a cross compiler. A cross compiler is simply a compiler that compiles (and links) but the output executable is based on the instruction set of the target platform.

This is exactly the situation that we will find ourselves in while trying to create our own personal devices with an Arduino or possibly building a project and using a Raspberry Pi as the brain.

Normally the process of software development is to create something, compile it and deploy it to the target platform. Unless you always know what you want and can code it perfectly each time the development flow would involve continuously performing these steps.

If you are planning on creating a device using an Atmel, the micro-controller family used in the Arduino, you would need to look up a few parameters to make sure that your instruction set is correct. The actual compiling of the source code is just a few lines.

```
avr-gcc -Wall -Os -DF_CPU=1000000 -mmcu=attiny85 -c blink.c -o blink.o
avr-gcc -Wall -Os -DF_CPU=1000000 -mmcu=attiny85 -o blink.elf blink.o
```

It isn't that I cannot type these commands from the command prompt but after doing it once or twice it does become tedious as well as being error prone. These commands could simply be put into a small script for building executables or transferring files. This solution is ok but not ideal for small projects.

The good news is that in the annals of time other programmers have also faced this situation of tedious and error prone and have come up with tools to help eliminate these problems. There is no limit to the number and types of tools involved in the development process but the one that will help automate the process of tedious commands for building software – make.

Installing gcc

> Linux
> sudo apt-get install gcc-avr binutils-avr avr-libc

Windows
https://gcc.gnu.org/

Command line build tools
The make command actually reads configuration files, called Makefiles, which contain the rules for generating programs. The make program determines which source files of a large program that may need to be recompiled. Each time the make program is run it will compare the dates of the files to see if any source files need to be recompiled because one or more of their dependencies have been changed.

It might sound odd to learn that the make command doesn't actually know anything about how to create software, it can simply analyze when changes have occurred and which steps need to be executed to rebuild a single file or even hundreds of files if necessary to rebuild a target.

If the make command does not have any specific knowledge of how to create programs then how can it possibly create programs or objects?

The make command is the ultimate in flexibility. When setting up the configuration file, which is known as a makefile, you can define your own commands.

```
e.g.
COMPILER = gcc
```

The only catch is that either the command, the c compiler in this example must be available in the users path or must be described with the complete path to the executable.

The make command is actually perfect when doing a lot of command line activities and it works as well for large programs as for small. In essence the entire process can be boiled down to a single template.

```
Target: dependencies
        command 1
        command 2
        command n
```

The dependencies is a list of objects such as header files, object files, or even libraries. The make command will look to see if the target exists and that the date and time of that object is more current than that of the dependencies.

```
Hello.o: Hello.h
        gcc -c hello.c -o hello.o

Hello: Hello.o
        gcc -o hello hello.o
```

The make utility will go through all the targets defined in the makefile and is clever enough to determine how the dependencies relate to each other and build any missing objects before building the programs that depend on them.

At its heart this is everything that the make utility can do. However, if you think about it the make utility must somehow do more than this. It should be obvious that the Linux Kernel doesn't have sixty thousand sets of explicit targets defined - one for each source file. The make command does have some very clever syntax that will let you create rules that will compile all files with a given extension.

The make command is very extensive and beyond the scope of this book but enough of the make command will be described or provided as examples that it can be used for your projects.

Installing make

>   Linux
>   sudo apt-get install make

>   Windows
>   https://www.gnu.org/software/make/

Data transfer

Using a cross compiler it is possible to create the actual program for a micro-controller but there is still the task of getting the program from your development PC to the micro-controller.

Essentially the way to get a program over to a micro-controller is to connect a wire between the computer and the micro-controller. This actually sounds pretty simple until you consider that a typical micro-controller will be a tiny chip with between 4 and 20 or more pins.

Most computers will have a USB port which actually is very convenient in the PC world but less so with small electronic parts. Yet, the solution to this problem is to add some electronics that can connect both the computer on one end to the micro-controller on the other.

Illustration 78: AVR Dragon with ZIF socket

Illustration 79: USBTinyISP

Yet, this small "hardware bridge" doesn't work in a vacuum, you also need software to use this AVR programmer to load the program into the destination chip. The name of the utility to do this transfer is Avrdude.

Avrdude is a program for uploading and downloading programs to the Atmel AVR family of micro-controllers. It can read several different input formats (Intel HEX, Motorola S-Record or binary) when transferring programs from your personal computer over to the micro-controller.

Avrdude transfers the program via communications port to the micro-controller. Well, the AVR programmer used to be connected to a serial port on the pc. Over the years the connection to the serial port was replaced by a USB connection.

In addition to transferring the program code over it can also set the fuse bytes on the micro-controller. Fuse bytes are special configuration for the micro-controller which is described further in the appendix "Describing an Atmel micro-controller".

Because the Avrdude is just another command line program which it is easy to add this to a makefile or a script when automating the development process.

Installing avrdude
    Linux
    sudo apt-get install avrdude

    Windows
    https://sourceforge.net/projects/avrdudegui/

Arduino IDE
Arduino IDE, just like most integrated development environments, is a software suite that allows you to perform all the steps in the developmental process. You can write your code, compile it, run it, debug it and follow your output using the serial monitor.

Features
- Compiler
- Data Transfer
- Editor
- Examples
- Manage dependencies

This development environment supports multiple different in-circuit serial programmers. Not only that this IDE can be used to flash multiple different target devices. If for some reason your device isn't already supported or you wish to upload new or modified firmware it is possible.

Arduino was designed with its own special boot loader. When creating a new project the editor will automatically create the file with two special methods.

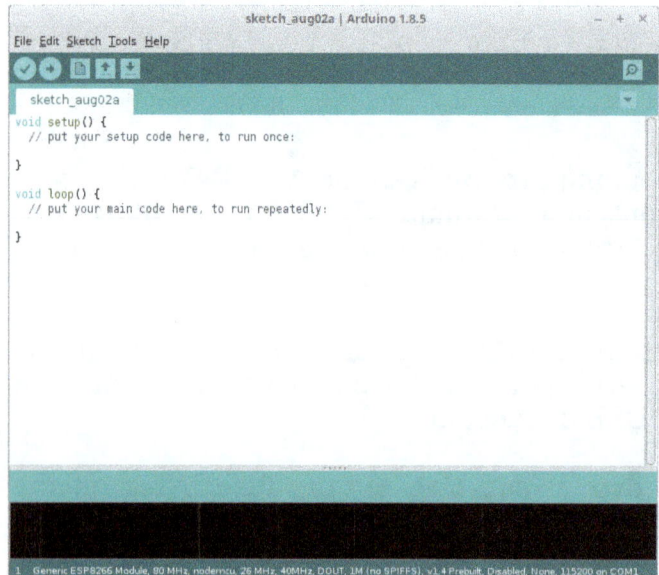
Illustration 80: Arduino integrated development environment with open sketch

This is done to clearly delineate which method needs to contain the initialization (the setup method), and which should contain the main logic.

Installing Arduino

Linux
https://www.Arduino.cc/en/Main/Software

Windows
https://www.Arduino.cc/en/Main/Software

Eclipse IDE
The Arduino IDE is an integrated development environments that ties together the task of editing code, compiling it to programs and it also lets you following along the debug output. The Arduino IDE actually does all of this but is geared specifically towards development for micro-controllers.

Eclipse is just another integrated development environment but it is flexible. In this case flexible means that it is possible to re-configure the IDE itself to support many different languages and uses. This is made possible as the base IDE is configurable using different plug-ins.

This development environment was originally developed for Java but in the meantime the Eclipse has different pre-configured releases.

- Java
- Java EE
- C/C++
- Android
- Java script & Web developers

- PHP
- Rust
- Parallel
- RCP & RAP
- Scout
- IOT
- Python
- Ogee
- Cloud IDE

If this list wasn't long enough, it is possible to take a clean version of Eclipse and configure it in a new and unique setup.

Installing Eclipse

Linux
https://www.eclipse.org/downloads/
https://www.eclipse.org/projects/

Windows
https://www.eclipse.org/downloads/
https://www.eclipse.org/projects/

6 ARDUINO

What ia an Arduino? An Arduino can simply be viewed as a AVR development board but what is that exactly. The AVR micro-controller architecture was developed at Nordic VLSI in 1996 by Alf Egil Bogen and Vegard Wollan. It is based on the Harvard micro-controller architecture. This technology was later sold to Atmel corporation.

Each of these AVR micro-controllers have quite a few features in common. The most important is that they have a varying number of input/output ports. Each of these ports have eight physical pins[25] that are accessible.

These pins can be configured to be either input or output. Other features of these micro-controllers are timers, hardware supported PWM, Two wire interface[26], UART which can be used for RS232 communication and Serial Peripheral Interface (SPI).

Is an Arduino simply a micro-controller on a board? Yes and no. The Arduino is indeed a micro-controller development board but it is also the entire development environment around it. This is both the software libraries that make it easy for the Arduino to use the UART to communicate with the outside world but also the configurable integrated development environment. This IDE makes it easy to cross compile the code and then transfer it over to the micro-controller.

This is what an Arduino is but the Arduino, just like most things, changes over time. The Arduino has changed quite a bit since it's first was introduced in 2004.

Arduino Uno	ATmega328
Arduino Duemilanove	ATmega168
Arduino Leonardo	ATmega32u4
Arduino Mega	ATmega2560

Most of these changes are due to the Atmel micro-controller powering the Arduino. The most obvious changes are the number of pins, speed and memory available but other functional changes also have occurred over the years.

Back in the day communication with the Arduino was using a serial cable but over the years the serial port has fallen by the wayside and has been replaced with the humble USB

---

25 The smaller AVR's which only have eight pins in total have fewer physical pins available.

26 The two wire interface is the Atmel equivalent to the I2C interface.

cable. Small but visible changes in how the board is flashed (e.g reset button). Other changes have been due to the advancement of technology. This is both due to larger more capable micro-controllers as well as surface mount versions which make it possible for creating much smaller Arduinos.

On the Arduino site there are over 25 different possible models[27] as well as a few shields for sale with at least that many older discontinued[28] models. This just shows how active the development of the Arduino is which is a much smaller view than all of the compatible versions that are constantly showing up on the market[29].

What does the Arduino offer that a small computer such as the Raspberry Pi cannot offer? The difference is that the Arduino is not a computer but a micro-processor on a board. This means limited memory, less expandability, slower speeds and no peripherals but it also means that there is no operating system. This does limit the kinds of tasks that this is suitable for.

The lack of operating system makes it a possible choice in situations where extremely accurate timings are required, in short tasks run on a micro-controller are deterministic. A normal operating system may seem responsive to the person sitting in front of it but there are interrupts that might prevent time sensitive measurements or actions from occurring in a timely manner. Another positive feature of no operating system is that the micro-controller can be simply turned off. Unlike a standard computer which needs to be shutdown in an orderly manner.

Micro-controllers, because of their cost and size, tend to be dedicated to a single task which further ensures that time sensitive tasks occur on time. Implantable medical devices or hard disk controllers are just two examples.

Every situation is unique and the flexibility of a computer is not always necessary or even desired. The Arduino has been around for a number of years and if you do have need of one there is certainly one for you. Simply select a Arduino that most closely fits all the requirements for its intended task.

	Arduino Uno	**Arduino Mega 2560**
Chip	Atmega 328	Atmega 2560
Power	5V	7-12v
Max current	40mA	50mA
Current per pin	-	20mA
Clock freq	16 mhz	16 mhz
Flash memory	32k	256k
usb	Standard	standard
Sram	2k	8k
PWM	6	15
Analog pins	6	16
EEProm	1k	4k
Interrupts	2	6
I2C1	Yes	yes

27 https://store.arduino.cc/

28 https://www.arduino.cc/en/Main/Boards

29 https://en.wikipedia.org/wiki/List_of_Arduino_boards_and_compatible_systems

SPI	yes	yes
Dimensions	68.6mm × 53.3mm	101.6mm x 54.45mm

Shields

The Arduino is an amazing platform for building electronic projects. One of the reasons is quite analogous to writing software. It is possible to bring in extra "functionality" into your project in a standardized way. In the case of hardware extra's these parts can be a individual chip or sensor.

This could be adding temperature, pressure or vibration sensor for measuring or project tool. You could also add a battery backed date and time module or a xy-axis joystick module.

The flexibility of adding some sensors is great but depending on what extra parts you need for your project might make the resulting solution awkward with one or more little sensors hanging by a wire.

The solution for this problem was solved by a small addon board called a "shield". The shield is a printed circuit board usually in the same size and shape of the Arduino which would fit on top of the Arduino. It would be connected by header pins to the Arduino this gives it a solid connection but makes the extension looks and behave as if it were a standard part of the Arduino.

The good news is not only can you attach a shield to your Arduino, with the right header pins you can stack multiple shields in order to get multiple sets of functionality. This can allow you to add both an ethernet port and a LCD screen. If you don't need this you may choose WiFi, Bluetooth or perhaps something even more exotic.

Yet this solution of adding shields does not offer endless expansion when for adding new functionality. The Arduino itself receives only 500ma of power. Any new shields cannot use more power than the Arduino makes available and some types of shields (Ethernet, WiFi, LCD with back light) use a considerable amount of power just for that one task. Power is not the only obstacle when adding shields to an Arduino. Each board needs certain pins to be available[30] in order to provide their functionality. Sometimes these pins can be re-defined in software but other times they cannot and when two or more shields attempt to use the same pins problems occur.

When problems occur, if not clearly defined in the documentation, you will need to read through the schematics in order to try and determine where the problem lies. The problem might not always lie in the hardware but it could also be in to the libraries that support them. The problem might be due to conflicts between the interrupts or because of how much flash or SRAM is being used.

Setup

The Setup of this device is a bit misleading when compared to the Raspberry Pi. The Arduino does not have an operating System, external storage or even a Video Adapter and because of this it is pretty easy to Setup. The set of instructions could probably look somewhat similar to the following.

> Take out of box
> Connect to power

[30] Available means that one pin is not being simultaneously used by two shields.

This is, of course, a more than slightly misleading. You can power up a brand new Arduino but it won't do anything other than convert power into a slight bit of heat. Without specific programming this little micro-controller board won't do anything useful.

Thus rather than discussing how to setup the Arduino it is more instructive to look at how to prepare the development environment. The installation of the Arduino IDE is both machine specific and quite easy. Just download the most current version of the IDE from the Arduino homepage and install it on your computer.

>   https://www.Arduino.cc/en/Main/Software

The installation is of the Arduino IDE is not significantly different than installing any other software on either Linux or Windows. The current version of the integrated development environment contains setup for all existing AVR boards.

This is perfect if you own or acquire a new Arduino and wish to use the IDE. One of the reasons why another, Eclipse, is so popular is because it is flexible enough that it can be configured to support different languages and tools.

A very similar level of flexibility is possible with the Arduino IDE. It is not possible to change which languages or tools but it is possible add support for new Arduino models. This is done through the concept of "cores".

An Arduino "core" provides a central place for all built in functions source files. These files are a form of abstraction. Thus with the abstraction of the core it is possible to add the support for non-Arduino boards. Creating such a core is non-trivial[31] but if you do create your own Arduino compatible device it is possible to get the same level of support from the Arduino integrated development environment. The IDE provides a "boards manager" option which can be used to actually install a new core and thus add new boards to the IDE.

Development IDE
The development environment encompasses all the tools that will be required for development and deployment. The most obvious tool is the compiler but this list also includes the utilities that are used for transferring the program to the Arduino. One of the nifty things about the Arduino is that it already contains a bootloader. The bootloader is the firmware that takes the place of an external programmer for uploading the program into the micro-controller. Without this functionality or if you were simply programming a Atmel chip you would need an external programmer, illustration 81, which would take the code and would upload it into the micro-controller

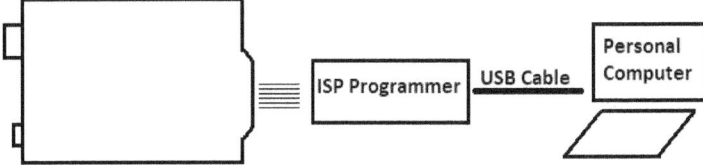

Illustration 81: Micro controller device with ISP programmer

---

31 https://atadiat.com/en/e-arduino-core-source-files-make-new-core-building-steps/

Illustration 82 shows how the Arduino is connected to the PC with a USB cable and transfer the program.

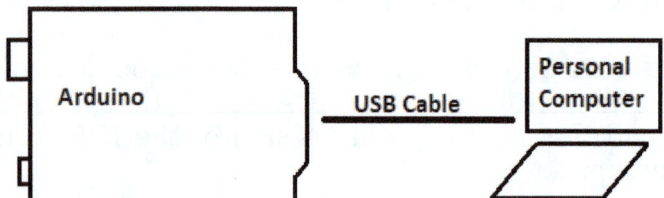

Illustration 82: Arduino connected with USB cable

The Arduino forces a certain form of discipline on the developers using it. In a manner similar to object oriented languages where the initialization is performed in a separate step before the main code of the micro-controller is run. This initialization method must exist and be named "init" and it is run once prior to calling the main body. The init method where one time configuration for the chip is done. This might be setting pins for input or output, setting up common memory structures, or even initializing communication with other shields or modules.

This separation of functionality is just good practice and is not really a burden on the developer. An initialization methods exist but this does not mean that there is a corresponding cleanup function defined. The reason for this is that a micro-controller is somewhat different than most computer programs. Most computer programs from web browser to word processor are started and eventually quit. The micro-controller never stops running.

The main code is actually an endless loop not dissimilar to the following pseudo code.

```
e.g.
void main()
{
        init();
        while (true)
        {
                loop();
        }
}
```

The Arduino IDE automatically support the separation between initialization and main program loop. Each time a new sketch, what the Arduino calls a program, is created it will automatically contain an empty stub for both the init and the loop functions.

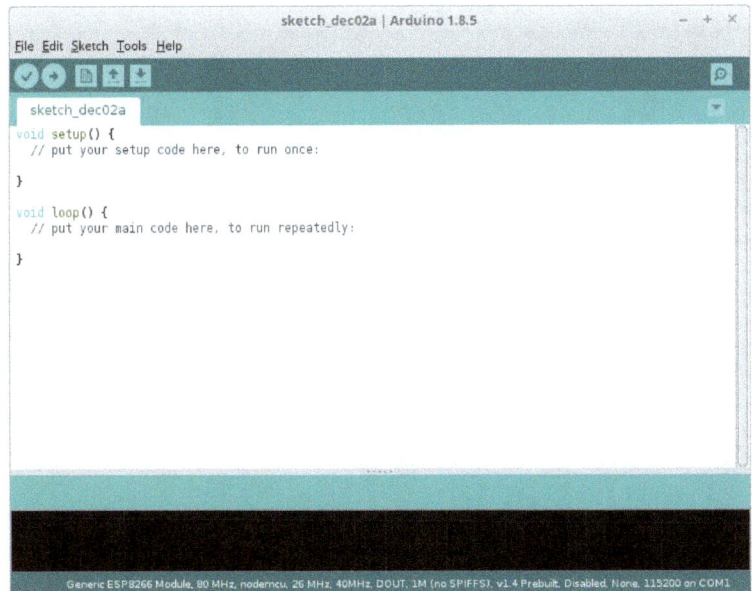

The IDE makes it trivial for compiling programs and uploading programs to the Arduino. Of course the menu structure does contain the options you might need but the IDE has a small toolbar with all the most used options.

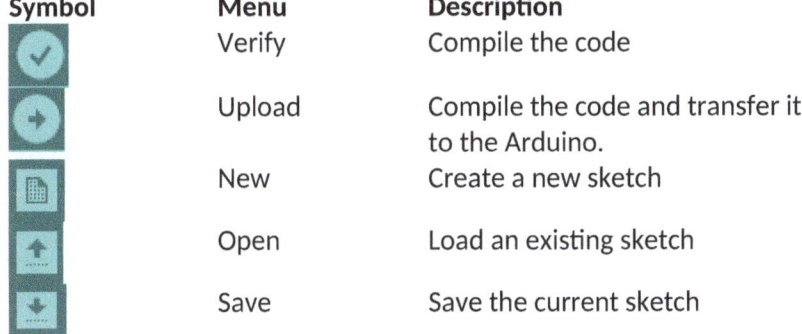

Symbol	Menu	Description
	Verify	Compile the code
	Upload	Compile the code and transfer it to the Arduino.
	New	Create a new sketch
	Open	Load an existing sketch
	Save	Save the current sketch

Most of these choices do exactly as the name implies. The verify option basically does a compile of the code while the upload will first do a compile and then transfer the resulting binary to the Arduino.

Programming

Development in the Arduino IDE is done using a subset of C/C++. Why a subset? Physical limitations of the device make some of these methods unnecessary. It is possible to have a printf function but the Arduino doesn't have a screen to output the result. In addition, memory is a very valuable commodity and linking in the printf functions would rapidly use it up.

The environment also has some special methods and functions designed specifically for supporting the Arduino itself. The Arduino has libraries full of functions to help make the process easier but before you can get started you need to become familiar with the hardware itself.

The Arduino Uno is powered by the Atmega328P microcontroller and although it is possible to use most pins for simple tasks some pins have extra functionality and you can only use those functions with those pins.

```
ATmega328P
        +--u--+
 PC6  1 |     | 28 PC5
 PD0  2 |     | 27 PC4
 PD1  3 |     | 26 PC3
 PD2  4 |     | 25 PC2
 PD3  5 |     | 24 PC1
 PD4  6 |     | 23 PC0
 VCC  7 |     | 22 GND
 GND  8 |     | 21 AREF
 PB6  9 |     | 20 AVCC
 PB7 10 |     | 19 PB5
 PD5 11 |     | 18 PB4
 PD6 12 |     | 17 PB3
 PD7 13 |     | 16 PB2
 PB0 14 |     | 15 PB1
        +-----+
```

Pin	Pin Name	Arduino Function
1	PC6	reset
2	PD0	Digital pin 0(RX)
3	PD1	Digital pin 1(TX)
4	PD2	Digital pin 2
5	PD3	Digital pin 3(PWM)
6	PD4	Digital pin 4
7	VCC	vcc
8	GND	ground
9	PB6	crystal
10	PB7	crystal
11	PD5	Digital pin 5
12	PD6	Digital pin 6
13	PD7	Digital pin 7
14	PB0	Digital pin 8
15	PB1	Digital pin 9
16	PB2	Digital pin 10
17	PB3	Digital pin 11
18	PB4	Digital pin 12
19	PB5	Digital pin 13
20	AVCC	vcc
21	AREF	Analog reference
22	GND	ground
23	PC0	Analog input 0
24	PC1	Analog input 1
25	PC2	Analog input 2
26	PC3	Analog input 3
27	PC4	Analog input 4
28	PC5	Analog input 5

When learning a new programming language you need to start with the basics. This is usually a program that is extremely simple, the actual lesson is not designed to do much more than go through the entire process of entering the code and performing any other necessary steps to create the desired output.

In quite a few cases this is the "Hello world" program which will display this test to the console or in a window.

In the Arduino world this program is usually called blinky and is designed to turn a single LED on and off. Once you are familiar with the Arduino and the integrated development environment it is not difficult to take a LED, a resistor and a breadboard and wire it all up so it can be turned on and off.

In order to spare the beginner the extra effort to setup it is possible to use a built in LED which already wired up. This LED is located near the transmit(TX) and receive(RX) LEDs.

```
#define LED 13                  // which pin

void setup()
{
pinMode(LED,OUTPUT);            // set the pin as output
}


void loop()
{
digitalWrite(LED,HIGH);         // turns led on
delay(1000);                    // sleep 1 second
digitalWrite(LED,LOW);          // turns led off
delay(1000);                    // sleep 1 second
}
```

Thinking in binary

Up until this point programming the Arduino has not really been explained. Just like any other computer it is possible to have variables and various control structures (for loop, while loop, if statement) but unlike a computer there is no screen for displaying the output and most actions are interacting with other hardware. A micro-controller board is more likely to be the brains and embedded into another larger device.

One such example would be in a smoke detector or smoke sensing service. When the signal goes from not sensing to sensing smoke the embedded device would then perform an action. This might be to activate a firefighting system which releases an insert gas such as halon or notifies the fire department of a problem.

Both of these are binary states. This works pretty well as the Arduino has quite a few digital pins as well as a few analog pins. The digital pins are used for enabling or disabling some external device. This is done by sending power out of the pin which is the equivalent of true and which is used by the recipient device. The converse yield the exact opposite functionality - no power usually prevents the receiving device from doing anything. The simplest case of this would be the recipient device being a LED. Obviously providing power will light a LED up and removing power will cause it to go dark.

It should also be understood that these pins can be used either for providing input, such as reading state of a button, or output to an external device.

However, digital pins are just one of two different types of pins. The function of analog pins may be more similar to how people experience the world around them. Not just black or white but also with shades of gray. The analog pins can actually accept values over a small range from analog devices. This analog value that is read in is interpreted by the Arduino as an 8 bit number.

The ability to read in analog values is a key feature when trying to make sense of input values from the physical world around you. It is necessary to interact with the physical world and the Arduino supports reading or writing out these types of analog values. However, there is a limited number of these pins on the Arduino.

This limitation is not a unique situation and it can be solved. The solution to that can be summed up with the acronym PWM. This acronym stands for pulse width modulation.

Thinking in analog with Pulse Width Modulation (PWM)
What is pulse width modulation? Pulse-width modulation is the process of changing a signal to have either a modified amplitude or modified width. The "width" is referring to one cycle where signal repeats. A single cycle contains both "on time" and the "off time" to represent the time that the signal is high or low respectively. The amount of time as a percentage that the signal is high relative to low is referred to as the duty cycle.

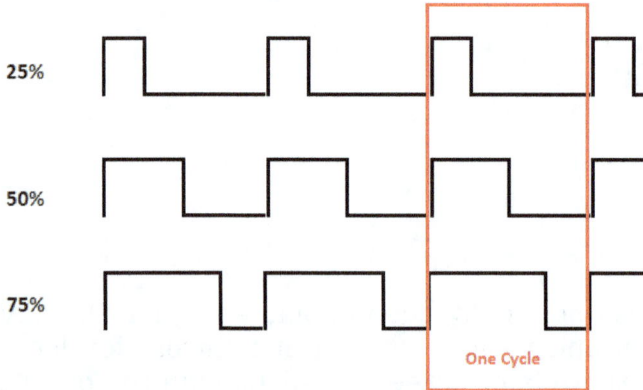

Illustration 83: Pulse width modulation showing several different duty cycles

When this is done rapidly the net effect is that the output will behave similarly to a constant voltage. What that "constant" voltage might be depends on the duty cycle. The two easiest cases to visualize what this is when the uptime is 100% or 0%. This can be calculated using both the "on time" and the voltage.

```
Ie.
On time * voltage        = average voltage
100% * 5V                = 5V
0% * 5V                  = 0V
```

Intuition isn't always correct in mathematics or physics but in this case, it seems reasonable if you have an up time of 50% then the average voltage might be similarly be one half of the original voltage. In this instance intuition is correct.

```
50% * 5V                 = 2.5V
```

How PWM works may or may not be interesting to the casual observer but how is this used in electronic circuits?

One of the answers is battery life. It is possible to use PWM to pass less power to a light emitting diode. In this particular case less power used means longer battery life but it also means a dimmer LED. How much dimmer depends on the duty cycle. A second use for pulse width modulation is dealing with direct current motors. Sending more power will make the motor turn faster while sending less will cause a slower spinning motor. Thus pulse width modulation can be used for gradual ramping up of a device as well as the gradual slowing of it.

However there is one last thing to consider for PWM. The duty cycle is the amount of time that the signal is high but this is one part of the equation. How often this happens per second or per minute also plays a part. This frequency is perhaps the most visible for LED's. If the power is on for 0.5 seconds and off for 0.5 seconds then the duty cycle of this is one hertz as this cycle happens once a second.

When a human looks at a LED it is obvious when the light is fully lit, dim or dark. The light would be full bright for ½ a second and dark for ½ a second. Most observers would probably say that this light is blinking.

This situation changes as this duty cycle becomes smaller and smaller. Ten times per second would still appear to be blinking fairly quickly but if that blink rate is 40, 50 or 60 times a second the human eye will start to blue these images together and we would see a dimmer light instead of a flashing light. It is this same technique that helps make cartoons moving pictures rather than a bunch of still pictures.

Pulse width modulation can be implemented in the way described - enable or disable the pin at a rate that supports the duty frequency that you need.

```
void loop() {
    digitalWrite(13, HIGH);      // turn the LED on
    delay(2);                    // wait for two milliseconds
    digitalWrite(13, LOW);       // turn the LED off
    delay(2);                    // wait for two milliseconds
}
```

This piece of code will have a duty cycle of 50% and a frequency of 250hertz. The technique is not difficult but to implement this in different parts of a program without affecting the rest of your code logic would be challenging without using some sort of timer driven method.

However there is a much simpler way to achieve this affect. The Arduino already has pins that support pulse width modulation. All that is needed is to define what the duty cycle should be for that pin. This is easy enough to do as each of these pins use a single byte for defining the duty cycle. You need to simply supply the value that when divided by 256 yields the duty cycle you are trying to achieve.

```
25      -> 25 / 256   -> 10%
64      -> 64 / 256   -> 25%
128     -> 128 / 256  -> 50%
```

Using the Arduino to control a LED is an example of communicating (reading/writing) directly with other hardware. Not all results are read from a sensor or module directly. Some of the modules use different communication protocols for transferring data or commands from the sensor to the Arduino.

Communication

It is possible to control simple components or modules simply by controlling the flow of power to them. This is quite primitive and it also makes it fairly difficult to transfer data between the different hardware modules. The good news is that this problem has already been solved in a number of different ways.

Inter-Integrated Circuit (I2C)
The I2C bus was developed in the 1980's by Philips Semiconductors. The purpose of this protocol was to provide an easy way to communicate between the CPU and associated peripheral chips in television sets. In the intervening years the I2C serial bus has extended its reach to desktops and laptops to interface with EEPROMS, specialized chips for monitoring temperature or power supply voltages.

It may not seem so revolutionary today yet it not only reduced the amount of wiring between components but also eliminates the need for additional logic for decoding the signals.

The solution utilizes only two wires for the data transfer, one is for the clock signal while the other is for the data. Utilizing this solution eases the effort to connect multiple chips together and saves money all at the same time. This protocol can support up to 127 individual devices on the I2C bus.

How exactly does I2C work
The data transfer via I2C is done by first sending a start condition. This is where the clock (SCL) is held high and the data line is brought low. The transfer of the data over the Serial Data line (SDA) is done after the start condition. The data is transferred by holding the SDA high or low for the minimum time in order for the slave device to see the bit. After the seven data bits have been transmitted they are followed by the direction bit. Once all eight bits have been transmitted the slave will transmit either an acknowledgment or negative acknowledgment. This will continue until all commands and data is transmitted. Once the transfer is finished a stop bit will be sent This is done by pulling both the data and the clock line high.

Below is an example of sending three bytes to a device with an address of 0111xx.

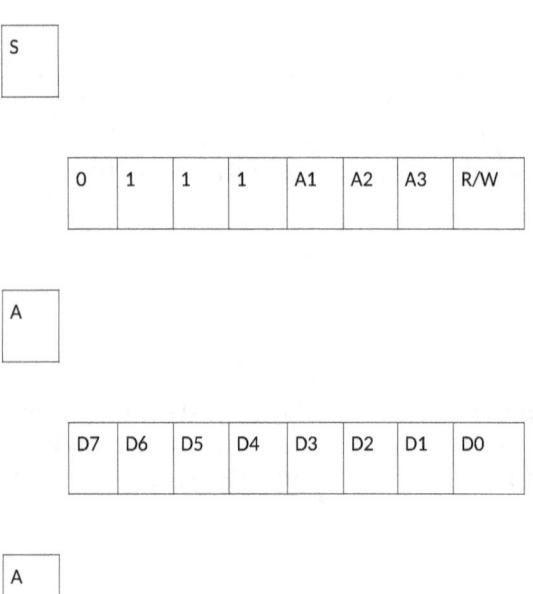

E7	E6	E5	E4	E3	E2	E1	E0

A

F7	F6	F5	F4	F3	F2	F1	F0

A

P

S	= Start bit
R/W	= Read / Write bit
A	= Acknowledgment
D0-7	= Data bits byte 1
E0-7	= Data bits byte 2
F0-7	= Data bits byte 3
P	= Stop bit
A1-3	= user defined address

Table 17: Example flow for transferring 3[32] bytes using I2C

Some integrated circuits have a hard coded address while others allow the I2C address to be defined in part by the manufacturer and in part by the user of the device. This can be accomplished in a number of different ways such as setting pins on the chip to either a high or low value.

The chip in the example, table 17, can have a user defined address by setting the lower three bits (A1,A2,A3) which when combined with the binary base address of 0111 gives a range of addresses. In this example the range would be 0x70 through 0x77. For this device it would be possible to have up to eight of the exact same chip on our I2C bus each with a different address.

To be comfortable with this protocol you may need to do additional research. There are most certainly books, data sheets from Philips and examples on the Internet[33]. Which method you select depends on your learning style.

This I2C protocol supports a number of different bus speeds ranging from the relatively slow standard mode bus speed of 100 kbits/s to high speed[34] mode of 3.2 Mbits/s. A very

---

32 This example appears to transfer four bytes but the first byte actually contains the address of the device receiving the data.

33 The tutorial which I found helpful was http://www.robot-electronics.co.uk/acatalog/I2C_Tutorial.html

clear explanation of the I2c protocol is available from NXP[35].

The inter-integrated circuit is not actually the only protocol that was created in an attempt to simplify circuitry.

Serial peripheral interface is a bus (SPI)
The Serial peripheral interface is a bus protocol that was developed by Motorola in the 1980's. If the comparison is limited to the fact that both I2C and SPI are bidirectional communication protocols then they are pretty similar.

The actual situation is actually a bit more subtle than that. The I2C solution uses two wires for the transfer data between devices while the SPI solution actually uses four wires.

> MOSI   Master out slave In (data output from master)
> MISO   Master in slave out (data output from slave)
> SCK   Serial clock (output from master)
> SS   Slave select

Wiring up of a single SPI device is not overly complex despite requiring four wires, it is only as the number of devices get added that things tend to become complex. All connected SPI devices will share the MOSI, MISO and SCK lines while each individual device will have its own slave select(SS) line. This slave select is what activates the device to start to listening.

There is also less democracy for this standard. For the SPI bus here is a single master device which can talk to any of the connected slaves yet the slave devices can talk only with the master device. Another difference is that there are no explicit addresses for each of the SPI slave devices nor is there an explicit transfer speed. It is possible to connect multiple devices each supporting a different speed as the common clock speed is lower than the maximum frequency for each of the devices involved.

The downside of the SPI protocol is that it has developed into a defacto standard. In general you need to check the datasheet of any devices that you will be using via the SPI bus. This is important as due to the lack of standardization it is possible to encounter a situation where some devices transfer the most significant bit first while others the least significant bit first.

Serial communications
Serial communication is just another protocol for transmitting data between two systems. The data is transmitted one bit at a time sequentially over a wire. This isn't the only way data can be transmitted, it could also be transmitted in parallel which would be to send the entire eight bits at the same time over eight wires.

Serial communications used to be a fairly common communication protocol for personal computers and peripherals. There where a lot of peripherals communicate with the PC using the serial protocol. The list starts with benign peripherals such as keyboards, mice and also included older devices such as modems and printers. These days a lot of those communications have been replaced by infrared, WiFi, Bluetooth or USB.

---

34 Newer standards of SMBUS support 1Mhz, 3.4Mhz with the latest version supporting 5Mhz.

35 http://www.nxp.com/documents/user_manual/UM10204.pdf

General serial communication can be actually pretty tricky due to the timing while transferring data but another advantage of the Arduino platform is the software environment that comes with it. This environment also includes serial communication libraries. The communication methods supplied make it as easy to either send or receive data over the serial channel between devices. All the heavy lifting is done under the covers so that transferring data is just a few lines of code.

```
void setup() {
  Serial.begin(57600);
}

// will be run continuously
void loop() {
  Serial.write("hello world");
  delay(1000);
}
```

This is the entire program! The variable "Serial" is a global variable from the Arduino libraries that will use the default serial UART hardware of the Arduino. Strictly speaking the Arduino Uno hardware only supports a single serial communication connection with hardware. It is possible to use other pins for serial communication supported by the Arduino libraries.

Some Arduino boards or more specifically the Atmel micro-controllers can support multiple hardware serial connections. It is possible to have multiple serial connections open at a time and copy data back and forth between them. The changes required to the above code only needs a few tiny changes to support serial communication for this same scenario.

```
SoftwareSerial mySerial(10, 11); // RX, TX

void setup() {
  // initialize default port
  Serial.begin(57600);
  mySerial.begin(57600);
}

void loop() {
  // run over and over
  if (mySerial.available()) {
    Serial.write(mySerial.read());
  }
}
```

This example will essentially forward input or output from the port defined by the variable mySerial to the default serial port.

The serial port is a neat way of either transferring data or commands between two different devices but there is a drawback. It is a powerful way to link to devices but it is point to point. It is not possible to have multiple devices directly communicate with each other.

When it is necessary for multiple devices to communicate with each other then perhaps either I2C or SPI might be a better solution.

Debugging

In general computers are very comfortable to program on because there are so many different methods for debugging a misbehaving program. When encountering a problem it can usually be traced quite simply by generating outputs in the form of dialog boxes or log files or can be done by debugging the program directly using an integrated development environment.

None of these "standard" methods of debugging are possible while creating projects on a micro-controller as there is no operating system, disk drive nor graphical environment. The solution for debugging on an Arduino is to send output over the serial connection to be monitored by the developer on the computer running the IDE. Not quite the same level of flexibility as on personal computers but this does allow for rudimentary debugging.

One of the downsides of using this serial monitor console is that the communication speed needs to be identical for both the development PC running the IDE and the device it connects to. This isn't a problem during development as the developer knows what speed the serial communication has been set to and can easily connect the console at that same speed. Should the speed not be identical for both the sender and receiver then the data is received to the communication port is garbled.

Summary
The Arduino is a great solution for low cost, computationally light tasks that need to be ready with the flick of a switch. The Arduino is small, has low power requirements and can accurately time events without the concern that the operating system will distort those measurements. The Arduino can be extended through the purchase of one or more shields to add additional functionality.

7 RASPBERRY PI

Model specifications

Raspberry Pi 3 Model B+

- Broadcom BCM2837B0, Cortex-A53 (ARMv8) 64-bit SoC @ 1.4GHz
- 1GB LPDDR2 SDRAM
- 2.4GHz and 5GHz IEEE 802.11.b/g/n/ac wireless LAN, Bluetooth 4.2, BLE
- Gigabit Ethernet over USB 2.0 (maximum throughput 300 Mbps)
- Extended 40-pin GPIO header
- Full-size HDMI
- 4 USB 2.0 ports
- CSI camera port for connecting a Raspberry Pi camera
- DSI display port for connecting a Raspberry Pi touchscreen display
- 4-pole stereo output and composite video port
- Micro SD port for loading your operating system and storing data
- 5V/2.5A DC power input
- Power-over-Ethernet (PoE) support (requires separate PoE HAT)

Setting up through first boot

Before getting started
In windows the administrator is the user with the necessary privileges for doing system maintenance. On the Raspberry Pi the administration user is the root user. The root user has no limitations to what actions it can perform. This user should only be used in situations that require such elevated permissions. It is possible to get these elevated permissions for a single command using the sudo command.

The sudo command is designed to temporarily take over the identity of another user to execute a command. The advantage is that the permissions will revert to the user's normal permissions after the command has finished running. It is even possible with some extra configuration to configure sudo that only certain users can use this command or even just run it for individual commands.

> Anyone can make a mistake, but it takes an administrator to make a real mess.
> - anonymous

Getting started
Just like any standard personal computer there are a few peripherals that are needed to make everything work, of course not all of them are absolutely required. This is true for the Raspberry Pi as it is just another computer only smaller.

Here is a list of the most common items which might be required by the typical pi user and some of them may already be sitting in the closet.

- Micro USB power supply
- HDMI monitor
- USB keyboard
- USB mouse
- HDMI cable
- 4 GB SD card
- Ethernet cable
- Powered USB hub
- Speakers
- Raspberry Pi case

All of these items are important in their own way, but make sure that you do get not only a micro USB phone charger but one that does indeed output at least 1 amp for the older Raspberry Pi's but the new Raspberry Pi 3 actually needs a high quality 2.5A micro USB.

It is not necessary to discuss each element from this list and for the most part it is the last three items on the list that are completely optional. A powered hub might be necessary when using a device that uses a lot of power such as an external hard disk or a printer. The Raspberry Pi 3B and newer computers have four USB ports and WiFi which takes some of the pressure off for needing a hub.

Getting started with the Raspberry Pi will require a SD card with the operating system. There are basically two ways to get such a card. Either purchase a SD card with the operating system pre-installed or prepare it yourself. There is nothing wrong with purchasing a pre-prepared SD card however due to how cheap SD Cards are and how easy they are to make you might want to create a SD Card with either different operating system versions or with each setup for a specific use.

Setting up the SD card
Preparing a SD card is slightly different depending on which PC platform you are currently using. I do not have an Apple Mac available, so I will only be providing explanations for Linux and Windows.

The first step would be to download the Raspberry Pi operating system image from the https://www.raspberrypi.org/software/operating-systems. On the official site there are a few different official operating system images which can be downloaded as well as several special purpose operating systems.

Any of these operating system images can be used to run your Pi, but some of theoptions are for a more specific purpose such as OSMC or LibreELEC. These operating systems are for creating your own home entertainment system and not suitable for experimenting with motors, LEDs and sensors. If you have a large enough budget it is possible to prepare multiple SD cards each setup for a different purpose. One of those could be to use the Raspberry Pi as your home entertainment system.

I would be remiss if I did not mention one of the official operating system distributions "New Out Of Box Software" (NOOBS). This version is intended to make the task of setting up a Raspberry Pi as painless as possible. It even has some support for switching the video mode in case you are not seeing any output on your display. The Noob image will help you to install Raspbian, LibreELEC, OSMC, or Windows 10 IoT Core as well as a few other OS images.

I have not had any problems with any of the Raspbian images that I have used. Thus, this is the OS image[36] that I will be using for my examples in the rest of the book.

The operating system image is not simply a file that will be copied to the SD Card but will overwrite the entire contents of the SD Card with a new files system and operating system. Downloading an image is just a few clicks from the web browser, what is there to describe? It isn't the downloading of the image that requires description but the validating of the image. There are two possible scenarios which end up badly and both can be prevented by validating the image.

The validation is important to verify that the image that has been downloaded was didn't experience any technical difficulties during the download not only that it is complete but not corrupted. If the file is incomplete or corrupted it won't work right, only if you are lucky will it fail catastrophically on boot, otherwise odd behavior or crashes may occur during normal operation.

A second reason to validate the image is to ensure that it is the same one from the Raspberry site and not another image with hidden malware installed in it.

The test for both situations is quite simple, simply run a program to calculate a checksum of the image and verify that sum against what is published on the Raspberry Pi site. The checksum is calculated with the SHA-1 algorithm. This algorithm will calculate a unique 160bit checksum from the file. The checksum must be calculated from the zip file containing the Raspbian image not from the unpacked image.

The steps for verification and card creation are as follows.

1. Calculate checksum
2. Unpack the file
3. Create card with OS image

Linux

Many utilities and applications are freely available in the open source world. This includes a lot of tools which have become standard for security or to verify data consistency. A Linux installation has all the tools that are necessary for verifying the image and creating the SD card should either be available on your installation or can be easily installed.

Calculate the checksum
The program sha256sum calculates the checksum for the file or files passed to it.

```
$ sha256sum 2018-11-13-raspbian-stretch.zip
a121652937ccde1c2583fe77d1caec407f2cd248327df2901e4716649ac9bc97  2018-11-13-raspbian-stretch.zip
```

---

[36] My examples should work with all of the different versions of the Raspian operation system but have been tested on the 2018-11-13-raspbian-stretch.zip release.

It is possible to compare each character visually, but it is also possible verify this without the manual effort. Simply pipe the output through the grep command using the checksum from the web site to see if the two checksums match.

> $ **sha256sum 2018-11-13-raspbian-stretch.zip | grep a121652937ccde1c2583fe77d1caec407f2cd248327df2901e4716649ac9bc97 | wc -l**
> 1

If the check sum matches the checksum then a one will be printed otherwise a zero will be printed.

Unpack the file
The unzip command will unpack the zip file. Simply run the program with the file as the command line parameter.

> **unzip 2018-11-13-raspbian-stretch.zip**

The actual operating system image file will be unpacked in the same directory as the original zip file.

Create card with OS image
Installation of the operating system image is done using the dd command. This can be one of the most dangerous commands to use if you are not paying enough attention it is possible to overwrite the incorrect partition which would destroy that partitions contents.

The following example although is valid on my computer may not be valid on other computers. The output device must be that of the card reader and this might be /dev/sdX or it might look similar to /dev/mmcblk. If you do not know the device for your card you will have to find this out first. This process is described in the appendix.

> e.g.
> dd bs=<size> if=<input file or device> of=<output file or device>
>
> dd bs=1M if=2018-11-13-raspbian-stretch.img of=/dev/sdc

It never hurts to run the sync command after some I/O command has been run as this will ensure that all the output buffers are flushed to the SD card.

> **sync**

Windows
In the past Windows did not have the same out of the box support for some utilities that are used in Linux to validate the integrity of the download or copy the image to the SD Card. This has changed since the release of Windows 7.

Calculate the checksum
The Windows tool, certutil.exe, is a command line program that can deal with certificates as well as calculate hash values.

The SHA1 hash value will be calculated by default but it is also possible to calculate MD2, MD4, MD5, SHA256 or SHA512.

[C:\](C:\) **certutil -hashfile 2018-11-13-raspbian-stretch SHA256**
SHA256 -Hash der Datei 2018-11-13-raspbian-stretch:
47 ef 1b 25 01 d0 e5 00 26 75 a5 0b 68 68 07 4e 69 3f 78 82 98 22 ee f6 4f 38 78 48 79 53 23 4d

CertUtil: -hashfile command completed successfully.

It is also possible to calculate the following hash values using powershell.

SHA1
SHA256
SHA384
SHA512
MD5

[c:\](c:\) **powershell -c "(Get-FileHash -a MD5 mysamplefile.iso).Hash"**
f99ca25bc9d1ddd1bb9f5f10cd90de48

However, this option only exists in PowerShell from version 4.0. PowerShell is a language so there are a number of alternate ways to calculate using PowerShell but I will not be elaborating on them here.

If you feel more comfortable you can find and download graphical tools to calculate these hash values from the Internet.

Create card with OS image
Windows also does not have a command similar to the Linux dd command for creating the SD Card, but again it is possible to download a program which can do the job. The program, Win32DiskManager.exe[37], can read and write image files from SD Cards or USB devices.

The GUI for this program is also quite simple. Pressing on the folder icon will bring up a standard windows search dialog to select the image file.

---

37 http://sourceforge.net/projects/win32diskimager/?source=dlp

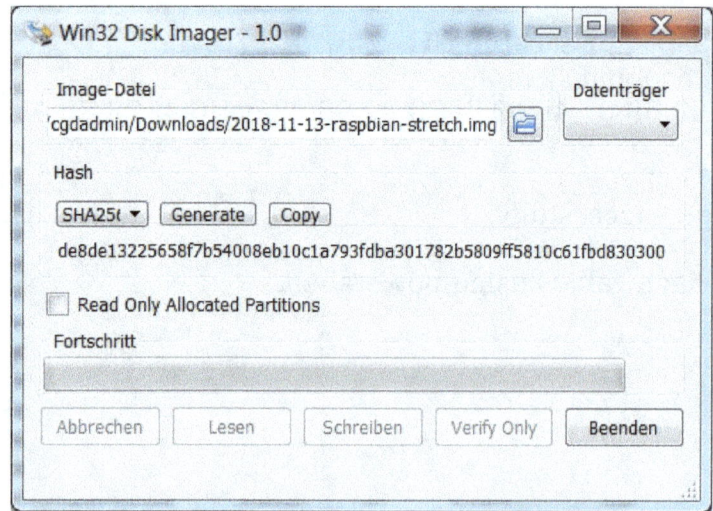
Illustration 84: Windows SD card image burner

The device drop-down box is a list of all drives that the program sees as valid USB or SD Card disk drives. Simply select the drive of the SD Card and press the Write button. The progress will be displayed on the progress bar and a window will pop up when the process is complete.

There is one peculiarity of this program, that is it expects that the SD Card is pre-formatted and that it only has one partition. Typically, if there is more than one partition on your SD Card you already know all about partitioning. I won't go into this further at this point but more information on this has been briefly described in the appendix.

First Boot (Raspbian Stretch Lite)
The first time the Raspberry Pi boots up the file system will be expanded to fill the rest of the SD card and then the Raspberry Pi will be rebooted. This minimal operating system will require you to log in using the following credentials.

    User   pi
    Password raspberry

There is one small catch for any Raspberry Pi enthusiasts who do not have a qwerty layout. The password does contain the letter "y" which on the German keyboard has been switched with the letter "z". The Raspberry Pi doesn't know you are not using a UK layout and so typing in raspberry will not work, you will need to use raspberrz as the password due to these letters changing places on the keyboard.

If your keyboard has any of these letters in different locations from the standard location on a Qwerty keyboard layout you may need to substitute other letters when entering the password.

Once you log in you will be able to start the configuration tool.

    $ sudo raspi-config

This will bring up a similar text windowing application where you can change your operating system configuration.

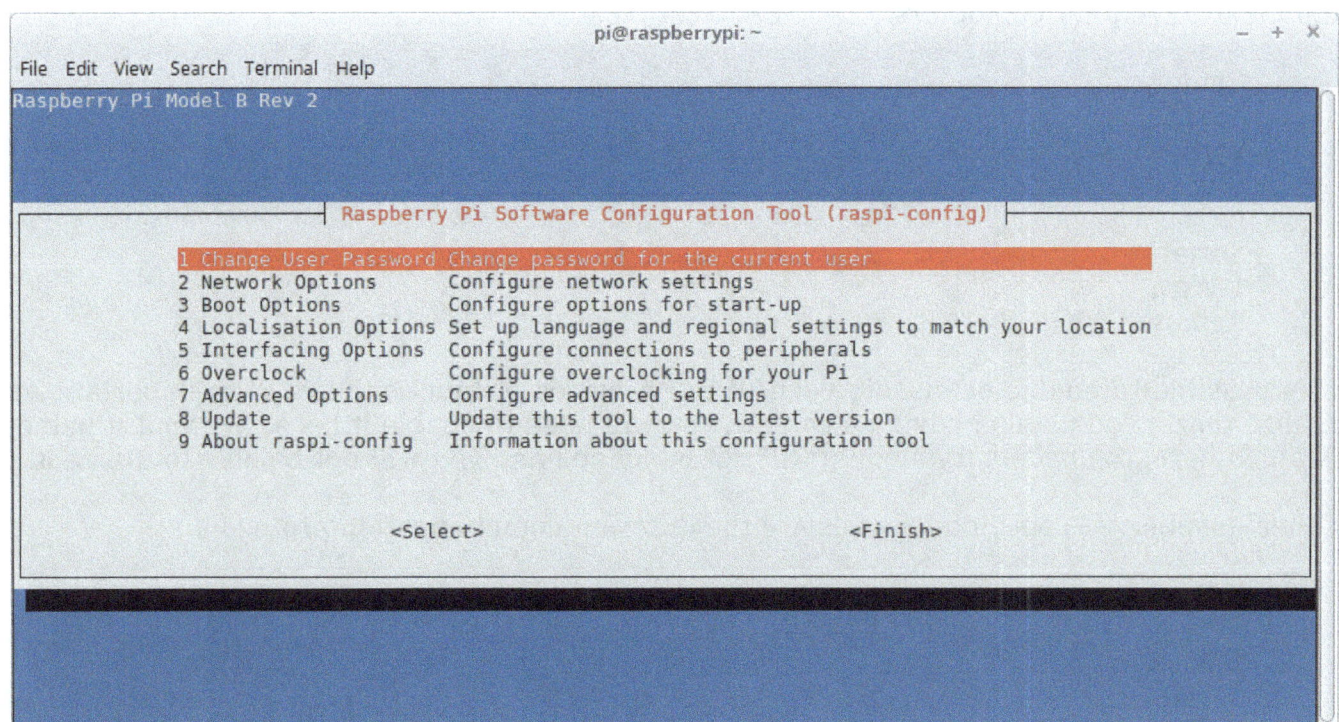

Illustration 85: Raspi-config program

Navigating through this application is done with using arrow keys and pressing the enter key to select the option.

It is not so productive to go through the description of each and every menu and sub-menu choice especially as most choices are well named and the functions that they perform are reasonably obvious. I will however describe a few menus that may be less obvious or have no corresponding personal computer equivalent.

The first menu choice is to change your password. This is neither new or unusual functionality, but it should be changed as soon as possible for security reasons. The reason is that the administrator user is "pi" and this password is initially set to "raspberry". This information is readily available on the internet which makes this an easy entry point to your machine and should be changed immediately.

Desktop / CLI
This configuration will setup the Raspberry Pi to either boot to a command line console or to a windowing environment. In both autologin cases the user will be "pi".

> Console
> Console autologin
> Desktop
> Desktop autologin

This option will let you decide how the Raspberry Pi will be booted up and whether or not you will be prompted for login credentials. You can pick which of these options best fits your work style.

Interfacing Options

This sub-menu contains a lot of choices for the functionality that makes the Raspberry Pi unique over regular computers.

> Camera
> SSH
> VNC
> SPI
> I2C
> Serial
> 1-Wire
> Remote GPIO

It is possible to enable or disable each of these pieces of functionality. It is important to realize that if you plan on having a camera you will need to enable it here. Without it being enabled here, the necessary module will not be loaded and you will not be able to utilize it.

Other Raspberry Pi specific features are the different communication protocols.

> I2C
> SPI
> 1-Wire
> Serial

Most of the small peripherals, add-on cards or Arduino devices can be communicated with from the Raspberry Pi using one of these four protocols. These options are all utilized from the GPIO pins.

In addition to allowing your Raspberry Pi to communicate with other attached devices it is also possible to connect to it from other computers. In order for that possibility you will need to enable SSH, VNC or both depending on how you plan on connecting.

Secure shell (SSH) will allow you to open up a shell window on the Raspberry Pi from another computer on your network. The connection will be encrypted so you can access this machine over the internet without the fear of your communications being found out. However, in order for all of that to work the ssh server needs to be started. This option determines if the ssh server is enabled to accept remote connections.

Not everyone enjoys only using a command prompt for checking on the status or running programs on network machines. The counterpoint to secure shell is Virtual Network Computing. When VNC is enabled it is possible to open up the desktop of the Raspberry Pi on the computer you are connecting from. This gives the user the more comfortable windowing environment with the same convenience as sitting in front of the actual Raspberry Pi.

More information on both of these techniques are described later in this chapter.

First Boot (Raspbian Stretch with desktop)

The first time the Raspberry Pi boots a program will be run which walk you through the configuration of the Raspberry Pi.

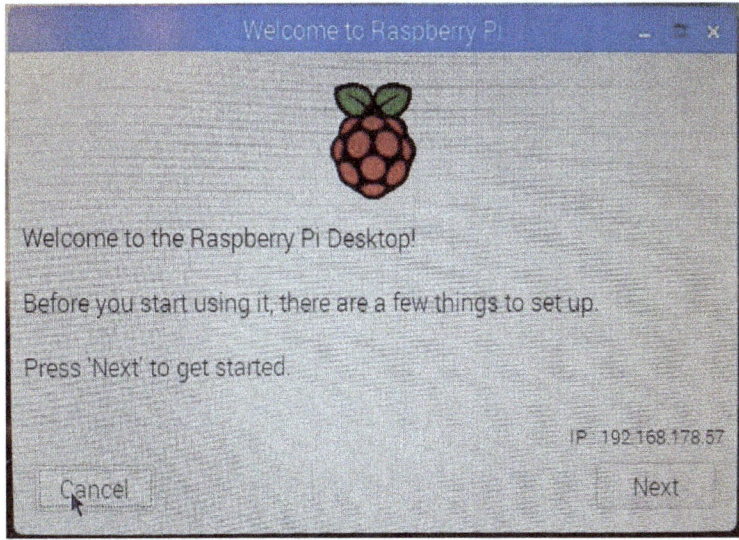

Illustration 86: First boot dialog

I have the Raspberry Pi connected to the network using the Ethernet port which is why I have been assigned the ip address – 192.168.178.57.

The next dialog is for setting up the environment. The dialog allows you to select which country, language and time zone. This is actually a really nice touch, but I did have a small problem. It isn't possible select some values that don't naturally match up. I wanted to select Germany as the country but then the language could only be a dialect of German. I also tried keeping the country as the United Kingdom but when this is selected the only time zone available is London.

The next dialog that opens allows you to select your SSID from a list of available wireless networks. Simply enter WiFi password and your networking configuration is all setup. The final prompt is to update the repository configuration on the Raspberry Pi.

Remote connections to Raspberry Pi
Computers without a monitor, mouse and keyboard are referred to as a headless computer. When the Raspberry Pi is your only computer then this option makes no sense. There would be no way to see or to program, but when more computers exist in a networking environment the flexibility becomes more apparent.

Secure shell
Headless computing may be done for cost reasons or perhaps even for space reasons. A headless computer is one that has neither keyboard, mouse or monitor. Typically, this model is used in situations where the server is setup and then it will not need much if any control by the administrators. Because the Raspberry Pi's operating system is Linux, it is possible to run multiple processes and have multiple terminal sessions simultaneously.

Using secure shell, it is possible to connect to the Raspberry Pi from Windows or Linux machine and open multiple text windows directly on the pi. These terminal windows behave similar to other command prompts which allows you to run programs and see their output. All programs or processes that are started from this window will be run on the Pi.

Windows users will need a secure shell[38] (SSH) client to connect to the Raspberry Pi. A very good SSH client is PuTTY[39]. Putty is another program which can be downloaded and run without any installation, which makes it a nice portable tool.

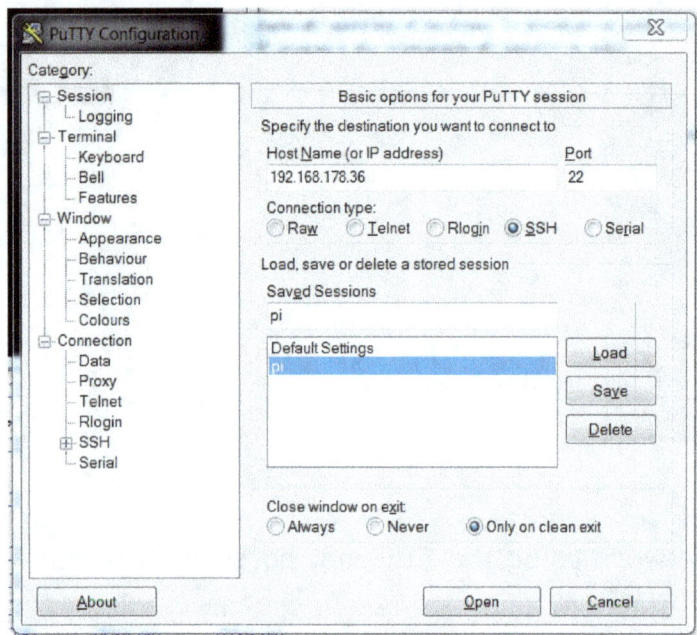

Illustration 87: Putty configuration

Putty will allow you to save the connection as well as other configuration options of frequently visited machines. It is as simple as entering enter the name of the machine or ip address but putty also allows you to customize the session with colors, window titles or other necessary configuration.

Linux users will most likely already have a SSH client installed on their machine. Connecting to another machine is as simple as the following command.

 ssh <user>@<machine name or ip address>

 ssh pi@192.168.178.57

However, you can only use putty or SSH when you know the machines IP address[40]. It is easy enough to find out the IP address of the Raspberry Pi by looking at the network setup on your Raspberry Pi. Rather than delving deeply into networking we will simply display the network interfaces that the Raspberry Pi is aware of. Simply run the ifconfig command from a terminal window.

 $ ifconfig

38 A secure shell is just a command prompt window on another computer where the communication has been encrypted.

39 PuTTY homepage is a collection of telnet, secure shell client, secure copy and other related associated utilities. http://www.chiark.greenend.org.uk/~sgtatham/putty

40 Not entirely correct, if you know the machine's name and the machine's name is in DNS, but that is beyond the scope of this chapter.

```
eth0: flags=4163<UP,BROADCAST,RUNNING,MULTICAST>  mtu 1500
        inet 192.168.178.57  netmask 255.255.255.0  broadcast 192.168.178.255
            inet6  2003:cb:173d:4900:990d:471:a9dd:cbe3   prefixlen 64   scopeid 0x0<global>
        inet6 fe80::cf0e:a3bd:d497:e8d4  prefixlen 64  scopeid 0x20<link>
        ether b8:27:eb:d5:c1:b7  txqueuelen 1000  (Ethernet)
        RX packets 782  bytes 89960 (87.8 KiB)
        RX errors 0  dropped 2  overruns 0  frame 0
        TX packets 326  bytes 82787 (80.8 KiB)
        TX errors 0  dropped 0 overruns 0  carrier 0  collisions 0

lo: flags=73<UP,LOOPBACK,RUNNING>  mtu 65536
        inet 127.0.0.1  netmask 255.0.0.0
        inet6 ::1  prefixlen 128  scopeid 0x10<host>
        loop  txqueuelen 1000  (Local Loopback)
        RX packets 0  bytes 0 (0.0 B)
        RX errors 0  dropped 0  overruns 0  frame 0
        TX packets 0  bytes 0 (0.0 B)
        TX errors 0  dropped 0 overruns 0  carrier 0  collisions 0
```

In this example the Raspberry Pi is connected using the Ethernet cable (eth0) and has the IP address of 192.168.178.57, and the normal loopback interface is also running. The loopback interface can be used to test networking software on the same computer.

It is too bold of a statement to say that networking always works, but I have not experienced any problems with it on my Raspberry Pi that was the fault of the Raspberry Pi or Raspbian. The Raspberry Pi will get it's IP address from the DHCP server on the router. If for some reason this is not happening, no IP address will be displayed when looking at the networking setup.

$ **ifconfig**
```
eth0: flags=4163<UP,BROADCAST,RUNNING,MULTICAST>  mtu 1500
        ether b8:27:eb:d5:c1:b7  txqueuelen 1000  (Ethernet)
        RX packets 782  bytes 89960 (87.8 KiB)
        RX errors 0  dropped 2  overruns 0  frame 0
        TX packets 326  bytes 82787 (80.8 KiB)
        TX errors 0  dropped 0 overruns 0  carrier 0  collisions 0

lo: flags=73<UP,LOOPBACK,RUNNING>  mtu 65536
        inet 127.0.0.1  netmask 255.0.0.0
        inet6 ::1  prefixlen 128  scopeid 0x10<host>
        loop  txqueuelen 1000  (Local Loopback)
        RX packets 0  bytes 0 (0.0 B)
        RX errors 0  dropped 0  overruns 0  frame 0
        TX packets 0  bytes 0 (0.0 B)
        TX errors 0  dropped 0 overruns 0  carrier 0  collisions 0
```

The output is virtually identical when compared the previous output except that the second line which displays the IP address is missing. This is an indicator of a "network" problem. This might be that the Ethernet cable is not plugged into either the router or to the Raspberry Pi, or it might be an indicator that the router is setup securely and isn't giving out ip addresses to unknown computers. If you believe that everything is setup correctly and you still do not have an ip address, see the troubling shooting section at the end of this chapter.

Virtual network computing (VNC)
Setting up VNC is actually only interesting when a desktop is involved and it is possible to set this up from the desktop with only a little bit of command line necessary. The VNC service listens on a specific port for the VNC client to connect to.

VNC Server setup
The virtual network computing process is just a server process, in this case on the Raspberry Pi, which forwards all video to the receiving client. The default port used is 5900.

1. Update the repository
sudo apt-get update

2. Install the vnc server
sudo apt-get install realvnc-vnc-server

3. Enable vnc in desktop
From the desktop the VNC server needs to be enabled. This is done from the Raspberry Pi Configuration submenu.

Menu → Preferences → Raspberry Pi Configuration → Interfaces

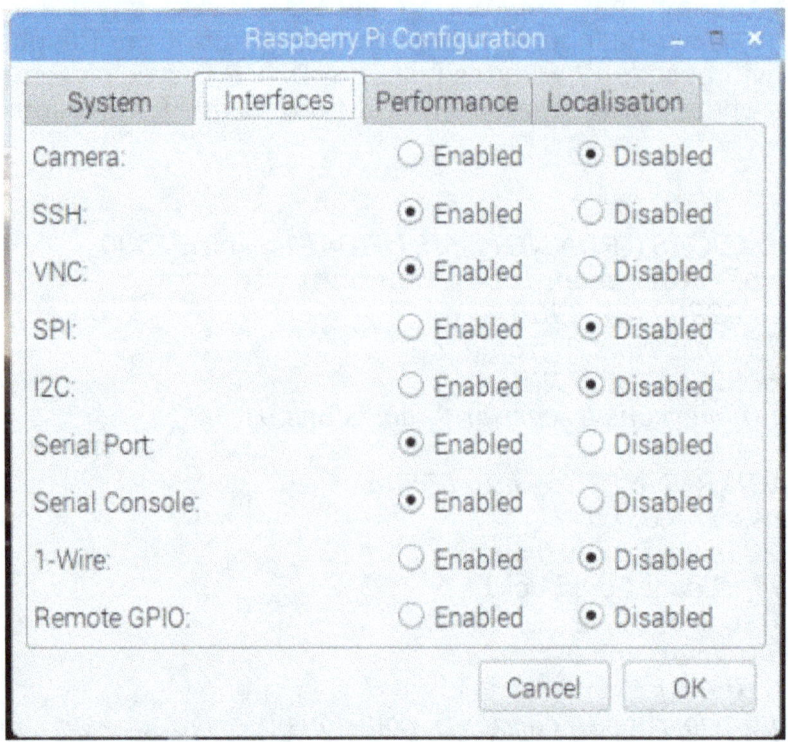

Illustration 88: Raspberry Pi GUI for setting up system daemons

Once all of this is done the Raspberry Pi is all set. All you need to do is to connect to it from somewhere else. However, having any connection to your server open does increase the risk that some bad person may connect and do bad things. Thus there are a few other commands you should know in order to disable this process.

All of these commands should be run from the command line. These are used to start, stop or change the runtime status on the next boot.

Description	Command
Starting server now	sudo systemctl start vncserver-x11-serviced.service
Starting server on next boot	sudo systemctl enable vncserver-x11-serviced.service
Stopping server now	sudo systemctl stop vncserver-x11-serviced.service
Disabling server on next boot	sudo systemctl disable vncserver-x11-serviced.service

VNC Client setup
There is actually a lot less to setup when connecting to a machine running VNC. The important steps are do you have a compliant VNC client and what port does the server expect to be connected to.

Just like any other type of software (editors, word processors) there are a number of different VNC servers and clients available. The VNC protocol both fairly simple and reasonably standard so you could expect just about any VNC client to connect. Any VNC client will not use any enhanced features that have been added to the VNC server protocol outside of the standard RFC 6143 which is for the transfer of rectangles of pixel data.

The good news is that it isn't necessary to download a number of different clients to see which one works as we can download the VNC client from RealVNC which was designed to work with this server.

> https://www.realvnc.com/en/connect/download/vnc/

Not only that but the client is available for the following platforms.

> Windows (32 and 64 bit)
> MacOS
> Linux (Debian pakcages, Red Hat packages, Generic tarball)
> Raspberry Pi
> Solaris
> HP-UX
> AIX

The VNC protocol, RFC 6143, refers to the transfer of a rectangle of pixel data. Because of this standard many VNC clients are compatible. Despite this limited compatibility the best choice would be to download the RealVNC client. It is perhaps folly to show the exact name and version number considering how quickly software is released but the current version of the client for Linux is **VNC-Viewer-6.19.107-Linux-x64**. This version has, as hoped, no problems connecting with the VNC server running on the Raspberry Pi.

Setting up WLAN
There are a few ways to setup a wlan network connection, either from the Raspberry Pi GUI using the WiFi Config program or the raspi-config configuration tool. Either of these methods of setup is ok, it simply depends on what you feel most comfortable with.

WiFi Config
From the desktop is a very nice WiFi setup program called WiFi Config. If you already know all the necessary information for connecting to your Wlan you can simply begin otherwise, there is even a scan option to locate the Wlan's in your vicinity, but you will still need to know the password.

To add a new wireless network, you will need to do the following steps.

> 1. click on manage networks tab
> 2. click on the add button
> 3. Enter the required information
>   - SSID
>   - Authentication method
>   - Encryption
>   - PSK (e.g password)
> 4. click on the add button
> 5. check status

At this point the wireless entry is saved, and if all the information was correct, a wireless connection is created. If you click back on the current status tab you can see the current status along with the assigned IP address.

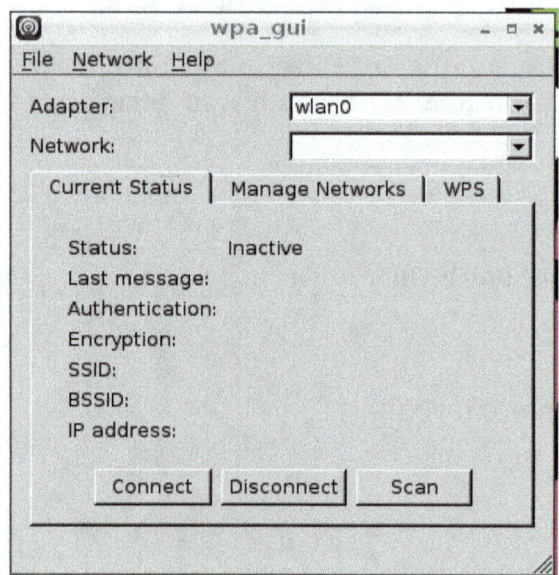

Illustration 89: Start WiFi configuration

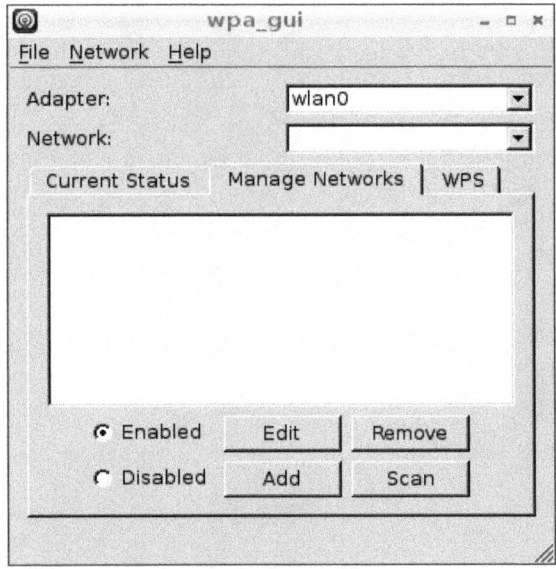

Illustration 90: Step #1

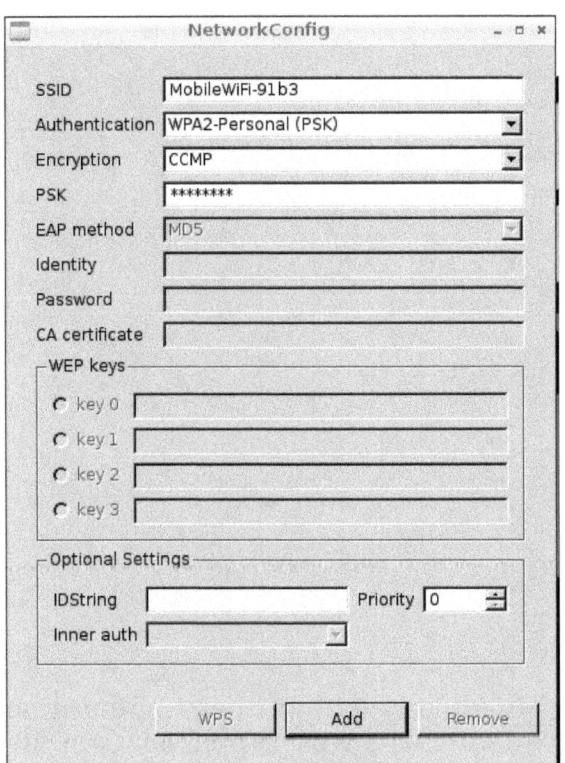

Illustration 91: Step #2 & Step 3

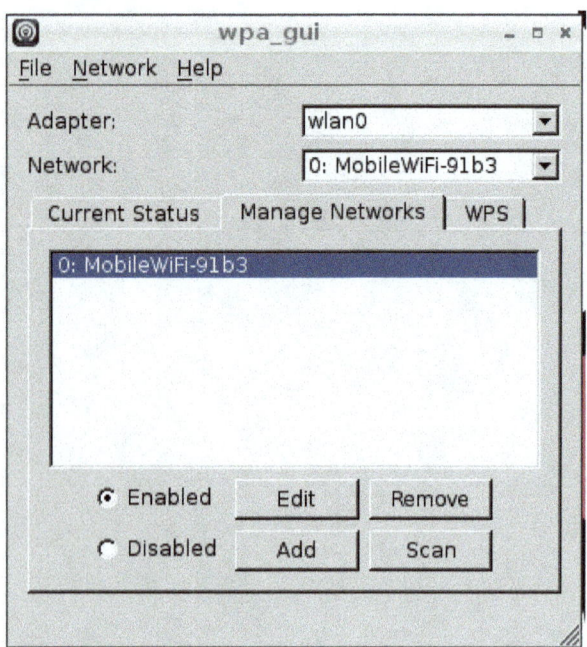

Illustration 92: Step #4

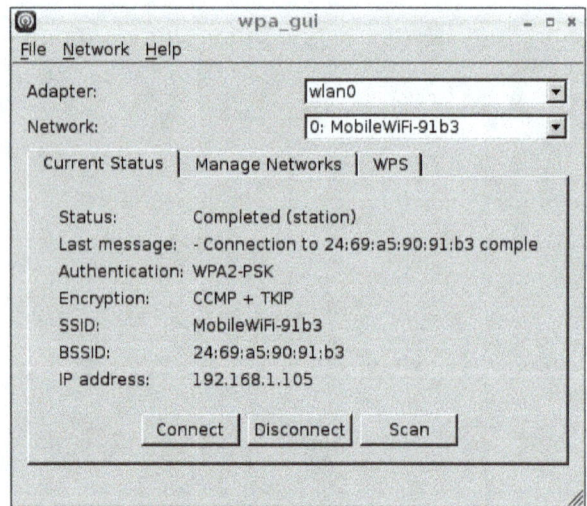

Illustration 93: Step #5, final setup

If you want to verify the values in another way you can actually look at the configuration file where they were written. The WiFi values are written into the wpa_supplicant.conf file in the /etc/wpa_supplicant directory.

Raspi-config
The raspi-config tool actually makes it easy to setup the WiFi. You select which country you are in and then you need to enter the SSID and the password.

Package manager and repository
Some years back the installation of programs to Linux was a bit more adhoc. Software was downloaded and installed on your system. Depending on the developer of that software package it may be a binary, source code with a make file or even source code and other custom scripts for building the program. It was up to the user to make sure that all the

required libraries were also installed. This was not a terribly fun process which could end up as trial and error depending how well the build process was documented.

Unsurprisingly, the community worked to come up with a better process. This improved method was the so called package manager. The package manager is a collection of tools which is used not only for the installation of new programs but also updating, removing and even upgrading the installation itself.

The package manager itself would install a software package. The software package would contain all the necessary information for installing the actual software and all of its related dependencies. In addition to the software and its dependencies these packages also contain other meta data such as version, description, author as well as other information to ensure package consistency such as checksums.

The package manager itself need not have a GUI front end, it depends on the actual distribution. Most distributions have both a command line package manager as well as a GUI front end for it.

Each package manager is configured to point to its repository. The package manager and the repository can be though of as the equivalent to an app store for modern day smart phone. What makes this paradigm even more powerful is that it is possible to configure the package manager so it can point to multiple different repositories.

This flexibility makes it possible to point a package manager to the developers repository for his or her program. That is important as software repositories may not always contain the current or even beta versions of a software package.

Command line

The Raspberry Pi has a GUI front end for the package manager but it is usually faster and easier to use the command line tool. This is not because some command line person (e.g me) feels more comfortable with it but this is the information that is typically available on the internet. It is much more precise to give short text name than to tell the user to search through the list of packages for an editor.

One of the primary preconditions of a package manager is a package repository. This command will update all of the lists that have been configured for the package manager. This command should be run prior to installing software.

    apt-get update

Installing software is not a typical task that a normal user can do, so it is necessary to either switch user to root or to use the sudo command to get elevated permissions during the install.

apt-get install <package>

    $ **sudo apt-get install dosbox**
    Reading package lists... Done
    Building dependency tree
    Reading state information... Done
    The following additional packages will be installed:
     libsdl-net1.2 libsdl-sound1.2
    The following NEW packages will be installed:

>   dosbox libsdl-net1.2 libsdl-sound1.2
> 0 upgraded, 3 newly installed, 0 to remove and 0 not upgraded.
> Need to get 781 kB of archives.
> After this operation, 2,427 kB of additional disk space will be used.
> Do you want to continue? [Y/n]

If indeed the requested package is not installed, you will be informed what additional packages or dependencies will be installed but the important thing is that you will be asked if this is what you really intended to do. The apt-get program will display everything that it is changing on your machine.

> Get:1 http://ftp.halifax.rwth-aachen.de/raspbian/raspbian buster/main armhf libsdl-net1.2 armhf 1.2.8-6 [14.1 kB]
> Get:2 http://ftp.halifax.rwth-aachen.de/raspbian/raspbian buster/main armhf libsdl-sound1.2 armhf 1.0.3-9 [75.5 kB]
> Get:3 http://ftp.halifax.rwth-aachen.de/raspbian/raspbian buster/main armhf dosbox armhf 0.74-2-3+deb10u1 [692 kB]
> Fetched 781 kB in 1s (760 kB/s)
> Selecting previously unselected package libsdl-net1.2:armhf.
> (Reading database ... 134951 files and directories currently installed.)
> Preparing to unpack .../libsdl-net1.2_1.2.8-6_armhf.deb ...
> Unpacking libsdl-net1.2:armhf (1.2.8-6) ...
> Selecting previously unselected package libsdl-sound1.2:armhf.
> Preparing to unpack .../libsdl-sound1.2_1.0.3-9_armhf.deb ...
> Unpacking libsdl-sound1.2:armhf (1.0.3-9) ...
> Selecting previously unselected package dosbox.
> Preparing to unpack .../dosbox_0.74-2-3+deb10u1_armhf.deb ...
> Unpacking dosbox (0.74-2-3+deb10u1) ...
> Setting up libsdl-sound1.2:armhf (1.0.3-9) ...
> Setting up libsdl-net1.2:armhf (1.2.8-6) ...
> Setting up dosbox (0.74-2-3+deb10u1) ...
> Processing triggers for mime-support (3.62) ...
> Processing triggers for hicolor-icon-theme (0.17-2) ...
> Processing triggers for gnome-menus (3.31.4-3) ...
> Processing triggers for libc-bin (2.28-10+rpi1) ...
> Processing triggers for man-db (2.8.5-2) ...
> Processing triggers for desktop-file-utils (0.23-4) ...

It is possible to install new software but it is also possible to remove it as well. This is somewhat misleading however.

> **apt-get remove <package>**
>
> **$ sudo apt-get remove dosbox**
> Reading package lists... Done
> Building dependency tree
> Reading state information... Done
> The following packages were automatically installed and are no longer required:
>   libsdl-net1.2 libsdl-sound1.2
> Use 'sudo apt autoremove' to remove them.
> The following packages will be REMOVED:
>   dosbox
> 0 upgraded, 0 newly installed, 1 to remove and 0 not upgraded.

After this operation, 2,213 kB disk space will be freed.
Do you want to continue? [Y/n]

Just like the install you are prompted if this is what you really wish to do. The misleading thing is that remove doesn't remove everything. This will remove the package but will not remove the configuration files that are on the system. If you really wish to remove everything then you need to use a different command.

apt-get purge

The purge option will remove both the package and the configuration files. This is usually what you want to do. It will allow you to "remove" the software and then install it truly fresh. Yet even that comes with a small caveat. Some software does store configuration data in your home directory and these configuration values will not be touched by either the "remove" or by the "purge" option.

Note: While upgrading to Kicad 5 from 4.7 I experienced some pretty odd results. In the end the problem was not with the software but actually that the 4.7 configuration was not removed. I manually removed the Kicad configuration in my home directory on my Linux laptop and the reinstall of Kicad 5 worked great.

Eventually you will want or need to update your software or Linux installation. There is an option for this as well.

apt-get upgrade

This option actually upgrades all the packages that are currently installed. This option is somewhat unique as it will upgrade the packages that are installed but it will not remove any installed packages nor will it install any previously uninstalled packages. This is a good option for being predictable as you know what was previously installed is just upgraded.

Sometimes when upgrading the packages in this way there can be problems after the upgrade is finished. One possibility of that would be if a package requires a new dependency that is not previously installed. This might be correctable by doing an upgrade which does not preserve the installed packages. This command is the upgrade-dist.

apt-get upgrade-dist

There might be a situation where you only want to upgrade a single package and this is also a possibility. You will need a slightly different syntax.

apt-get install --only-upgrade <package>

You may need to update packages from time to time on the Raspberry Pi but it may not be strictly necessary to do a distribution upgrade. If this ever comes to pass it would be quite easy to either create a new SDCard with a operating system or even just reuse the existing SDCard.

The apt-get command is used for installing, uninstalling and updating packages but there is another program which can be used to assist in the process. If you already have some idea what program you are looking for (e.g editor) but are not certain which editors are available. You can search through the list of available packages

apt-cache search <name or name part>

The apt-cache command will display packages with that name or name portion but this can produce a very large number of packages. That isn't to say that this program isn't useful but perhaps google might be a easier way to find out the package name of your software in your repository.

> $ **apt-cache search veracrypt**
> *libzulucrypt-dev - development files for libzulucrypt-1.2.0*
> *libzulucrypt-exe-dev - development files for the libzulucrypt-exe*
> *libzulucrypt-exe1.2.0 - provide the main functions of zulucrypt*
> *libzulucrypt-plugins - collection of plugins for zulucrypt*
> *libzulucrypt1.2.0 - provide the functions of zulumount*
> *libzulucryptpluginmanager-dev - development files for libzulucryptpluginmanager*
> *libzulucryptpluginmanager1.0.0 - provides support for plugins*
> *zulucrypt-cli - tool for encrypting volumes*
> *zulucrypt-gui - graphical front end for zulucrypt-cli*
> *zulumount-cli - tool that manages encrypted volumes*
> *zulumount-gui - graphical front end for zulumount-cli*
> *zulupolkit - handler the polkit privileges*
> *zulusafe-cli - cli that manages encrypted volumes*

Building software

Computer programs don't grow on trees, they are coded by regular people. The programs don't just magically end up in the software repository of the package manager. Someone goes through a considerable amount of work to compile the program and create a package that is in the proper form for the distribution.

Yet, different distributions have different package managers and different package managers sometimes have different package formats. One example of this is Debian and Red Hat. These two distributions have different package managers and despite both being Linux the packages from one is completely incompatible with the other.

Software developers who are part of either small teams or teams of one don't always bother to package up the program at all. Sometimes they make the code available as a simple tarball. This is a convenient way for software to be downloaded so it can be built on the target platform.

Usually the developer will include a Makefile, scripts or other method to make the build process possible for non-developers.

The simplest method is to have a configuration file that can be read by a program to build the program. This configuration file is called a Makefile and is used as input for the make command. This is a very convenient way to build a program but this can become tricky to have a single Makefile that will work for all platforms. Different platforms may have different tools and those tools may require different parameters.

A clever solution was developed to deal with this problem – the configure script. Quite a few multi-platform programs use this configure method for creating a build environment. This configure file is a script which interrogates the environment to gather a list of which programs are available for use. The configure script will then create a Makefile that will use the tools that are installed.

Build Example

To demonstrate how this entire process works I have downloaded a small project which builds a I2C library. This library allows C programs to utilize the I2C protocol to communicate with I2C devices.

```
$ mkdir workdir
$ cd workdir/
$ wget http://www.airspayce.com/mikem/bcm2835/bcm2835-1.60.tar.gz
```
--2019-10-01 21:49:19-- http://www.airspayce.com/mikem/bcm2835/bcm2835-1.60.tar.gz
Resolving www.airspayce.com (www.airspayce.com)... 192.185.48.187
Connecting to www.airspayce.com (www.airspayce.com)|192.185.48.187|:80... connected.
HTTP request sent, awaiting response... 200 OK
Length: 265906 (260K) [application/x-gzip]
Saving to: 'bcm2835-1.60.tar.gz'
bcm2835-1.60.tar.gz 100%[===================>] 259.67K 171KB/s in 1.5s
2019-10-01 21:49:21 (171 KB/s) - 'bcm2835-1.60.tar.gz' saved [265906/265906]

```
$ ls
```
bcm2835-1.60.tar.gz

```
$ gunzip bcm2835-1.60.tar.gz
$ tar xvf bcm2835-1.60.tar
```
bcm2835-1.60/
bcm2835-1.60/configure.ac
bcm2835-1.60/COPYING
bcm2835-1.60/examples/
bcm2835-1.60/examples/spi/
bcm2835-1.60/examples/spi/spi.c
bcm2835-1.60/examples/input/
bcm2835-1.60/examples/input/input.c
bcm2835-1.60/examples/spiram/
bcm2835-1.60/examples/spiram/spiram.h
bcm2835-1.60/examples/spiram/spiram_test.c
bcm2835-1.60/examples/spiram/spiram.c
bcm2835-1.60/examples/event/
bcm2835-1.60/examples/event/event.c
bcm2835-1.60/examples/pwm/
bcm2835-1.60/examples/pwm/pwm.c
bcm2835-1.60/examples/gpio/
bcm2835-1.60/examples/gpio/gpio.c
bcm2835-1.60/examples/blink/
bcm2835-1.60/examples/blink/blink.c
bcm2835-1.60/examples/spin/
bcm2835-1.60/examples/spin/spin.c
bcm2835-1.60/examples/i2c/
bcm2835-1.60/examples/i2c/i2c.c
bcm2835-1.60/config.guess
bcm2835-1.60/NEWS
bcm2835-1.60/depcomp
bcm2835-1.60/configure

```
bcm2835-1.60/aclocal.m4
bcm2835-1.60/compile
bcm2835-1.60/INSTALL
bcm2835-1.60/Makefile.in
bcm2835-1.60/missing
bcm2835-1.60/config.sub
bcm2835-1.60/Makefile.am
bcm2835-1.60/ChangeLog
bcm2835-1.60/README
bcm2835-1.60/install-sh
bcm2835-1.60/config.h.in
bcm2835-1.60/src/
bcm2835-1.60/src/test.c
bcm2835-1.60/src/bcm2835.h
bcm2835-1.60/src/bcm2835.c
bcm2835-1.60/src/Makefile.in
bcm2835-1.60/src/Makefile.am
bcm2835-1.60/ltmain.sh
bcm2835-1.60/AUTHORS
bcm2835-1.60/doc/
bcm2835-1.60/doc/Doxyfile.in
bcm2835-1.60/doc/Makefile.in
bcm2835-1.60/doc/Makefile.am
bcm2835-1.60/test-driver
```

$ cd bcm2835-1.60/
$ ./configure

checking for a BSD-compatible install... /usr/bin/install -c
checking whether build environment is sane... yes
checking for a thread-safe mkdir -p... /bin/mkdir -p
checking for gawk... no
checking for mawk... mawk
checking whether make sets $(MAKE)... yes
checking whether make supports nested variables... yes
checking whether make supports the include directive... yes (GNU style)
checking for gcc... gcc
checking whether the C compiler works... yes
checking for C compiler default output file name... a.out
checking for suffix of executables...
checking whether we are cross compiling... no
checking for suffix of object files... o
checking whether we are using the GNU C compiler... yes
checking whether gcc accepts -g... yes
checking for gcc option to accept ISO C89... none needed
checking whether gcc understands -c and -o together... yes
checking dependency style of gcc... gcc3
checking for clock_gettime in -lrt... yes
checking for doxygen... no
configure: WARNING: Doxygen not found - continuing without Doxygen support
checking for ranlib... ranlib
checking for gcc... (cached) gcc
checking whether we are using the GNU C compiler... (cached) yes

checking whether gcc accepts -g... (cached) yes
checking for gcc option to accept ISO C89... (cached) none needed
checking whether gcc understands -c and -o together... (cached) yes
checking dependency style of gcc... (cached) gcc3
checking that generated files are newer than configure... done
configure: creating ./config.status
config.status: creating Makefile
config.status: creating src/Makefile
config.status: creating doc/Makefile
config.status: creating config.h
config.status: executing depfiles commands

$ ls -ltr src
total 228
-rw-r--r-- 1 pi pi   205 Mar 8 2015 Makefile.am
-rw-r--r-- 1 pi pi   713 Jul 23 01:04 test.c
-rw-r--r-- 1 pi pi 32797 Jul 23 01:07 Makefile.in
-rw-r--r-- 1 pi pi 59146 Jul 23 01:10 bcm2835.c
-rw-r--r-- 1 pi pi 94133 Jul 23 01:24 bcm2835.h
-rw-r--r-- 1 pi pi 32433 Oct 1 21:50 Makefile

$ make
make all-recursive
make[1]: Entering directory '/home/pi/workdir/bcm2835-1.60'
Making all in src
make[2]: Entering directory '/home/pi/workdir/bcm2835-1.60/src'
*gcc -DHAVE_CONFIG_H -I. -I..   -g -O2 -MT bcm2835.o -MD -MP -MF
.deps/bcm2835.Tpo -c -o bcm2835.o bcm2835.c*
mv -f .deps/bcm2835.Tpo .deps/bcm2835.Po
rm -f libbcm2835.a
ar cru libbcm2835.a bcm2835.o
ar: `u' modifier ignored since `D' is the default (see `U')
ranlib libbcm2835.a
make[2]: Leaving directory '/home/pi/workdir/bcm2835-1.60/src'
Making all in doc
make[2]: Entering directory '/home/pi/workdir/bcm2835-1.60/doc'
make[2]: Nothing to be done for 'all'.
make[2]: Leaving directory '/home/pi/workdir/bcm2835-1.60/doc'
make[2]: Entering directory '/home/pi/workdir/bcm2835-1.60'
make[2]: Leaving directory '/home/pi/workdir/bcm2835-1.60'
make[1]: Leaving directory '/home/pi/workdir/bcm2835-1.60'

$ ls -ltr src
total 472
-rw-r--r-- 1 pi pi   205 Mar 8 2015 Makefile.am
-rw-r--r-- 1 pi pi   713 Jul 23 01:04 test.c
-rw-r--r-- 1 pi pi 32797 Jul 23 01:07 Makefile.in
-rw-r--r-- 1 pi pi 59146 Jul 23 01:10 bcm2835.c
-rw-r--r-- 1 pi pi 94133 Jul 23 01:24 bcm2835.h
-rw-r--r-- 1 pi pi 32433 Oct 1 21:50 Makefile
-rw-r--r-- 1 pi pi 121344 Oct 1 21:51 bcm2835.o
-rw-r--r-- 1 pi pi 123724 Oct 1 21:51 libbcm2835.a

Troubleshooting

Which version Linux
It is possible after some long time after installing Linux you may not remember all the details about the version installed. This information can be easily looked up from a few files on the installation.

> **$ cat /proc/version**
> *Linux version 4.14.98+ (dom@dom-XPS-13-9370) (gcc version 4.9.3 (crosstool-NG crosstool-ng-1.22.0-88-g8460611)) #1200 Tue Feb 12 20:11:02 GMT 2019*
>
> **$ cat /etc/os-release**
> *PRETTY_NAME="Raspbian GNU/Linux 9 (stretch)"*
> *NAME="Raspbian GNU/Linux"*
> *VERSION_ID="9"*
> *VERSION="9 (stretch)"*
> *ID=raspbian*
> *ID_LIKE=debian*
> *HOME_URL="http://www.raspbian.org/"*
> *SUPPORT_URL="http://www.raspbian.org/RaspbianForums"*
> *BUG_REPORT_URL="http://www.raspbian.org/RaspbianBugs"*
>
> **$ uname -a**
> *Linux raspberrypi 4.14.98+ #1200 Tue Feb 12 20:11:02 GMT 2019 armv6l GNU/Linux*
>
> **$ lsb_release -a**
> *No LSB modules are available.*
> *Distributor ID: Raspbian*
> *Description: Raspbian GNU/Linux 9.8 (stretch)*
> *Release: 9.8*
> *Codename: stretch*

Host identification changed

More specifically, receiving the following message when connecting to your pi using ssh.

> WARNING: REMOTE HOST IDENTIFICATION HAS CHANGED!

The problem with personal computers is that they are relatively big. It is simply not convenient to have multiple hard disks and to change which one you use depending on what games or programs you wish to use.

This is not the case with the Raspberry Pi. You can easily change which operating system, game or platform is loaded within minutes, all that is needed is a handful of SD cards. Each could have a completely different development environment or security setup.

Yet when trying to connect to the same Raspberry Pi while using different environments from another computer using your favorite terminal software you may see this message.

> @@@
> @   WARNING: REMOTE   HOST   IDENTIFICATION   HAS   CHANGED!

```
@@@@@@@@@@@@@@@@@@@@@@@@@@@@@@@@@@@@@@@@@@@
IT IS POSSIBLE THAT SOMEONE IS DOING SOMETHING NASTY!
Someone could be eavesdropping on you right now (man-in-the-middle attack)!
It is also possible that a host key has just been changed.
The fingerprint for the ECDSA key sent by the remote host is
30:55:63:87:13:54:b3:40:e0:97:cf:d9:38:01:d0:bd.
Please contact your system administrator.
Add correct host key in /home/cdock/.ssh/known_hosts to get rid of this message.
Offending ECDSA key in /home/cdock/.ssh/known_hosts:14
  remove with: ssh-keygen -f "/home/cdock/.ssh/known_hosts" -R 192.168.178.23
ECDSA host key for 192.168.178.23 has changed and you have requested strict checking.
Host key verification failed.
```

This message should be very worrying if you are connecting to systems on the internet or other computers that you know have not been changed in any way.

This message says the credentials that we have in our known_hosts file on line 14, differs from the host that we are connecting to. If indeed this happens to you during experimenting there is a very simple solution. The error message also comes with a solution to the problem.

ssh-keygen -f "/home/cdock/.ssh/known_hosts" -R 192.168.178.23

Running this command will remove the questionable entry but it will save the original file, just in case.

$ ssh-keygen -f "/home/cdock/.ssh/known_hosts" -R 192.168.178.23
/home/cdock/.ssh/known_hosts updated.
Original contents retained as /home/cdock/.ssh/known_hosts.old

The other solution would be to simply edit the known_hosts file in the .ssh directory and remove the offending line – line 14 in this example.

Keyboard Problems
To be honest, for most of my development on the Raspberry Pi I rarely used the graphical desktop. During my initial setup of the SD Card I had selected a generic 105 key German layout keyboard and the mapping was perfect – well without a desktop GUI.

Yet when using the graphical desktop I was surprised that the keyboard mapping seemed to be that of an American or English keyboard. For some reason the keyboard layout that I selected using the raspi-config program was not being used.

The fix that I did was a small change to the file "autostart" located in the /etc/xdg/lxsession/LXDE directory.

Before	After
@lxpanel --profile LXDE @pcmanfm --desktop --profile LXDE @xscreensaver -no-splash	@lxpanel --profile LXDE @pcmanfm --desktop --profile LXDE @xscreensaver -no-splash @setxkbmap -model pc105 -layout de

For anyone that may need a different keyboard layout then perhaps this or a small variation of this may work. Simply change the layout from "de" to your country code.

    e.g.
        fr   French
        pt   Portuguese
        en   English

System language

If for some reason you would like your operating system error messages in your own language make sure that you selected the proper locale.

Make sure that you change the locale under section of internationalization. You can check to see what locale is currently selected from a terminal window. Typing the locale command with no other options will display what the current settings are.

$ locale
locale: Cannot set LC_ALL to default locale: No such file or directory
LANG=en_GB.UTF-8
LANGUAGE=
LC_CTYPE="en_GB.UTF-8"
LC_NUMERIC=de_DE.UTF-8
LC_TIME=de_DE.UTF-8
LC_COLLATE="en_GB.UTF-8"
LC_MONETARY=de_DE.UTF-8
LC_MESSAGES="en_GB.UTF-8"
LC_PAPER=de_DE.UTF-8
LC_NAME=de_DE.UTF-8
LC_ADDRESS=de_DE.UTF-8
LC_TELEPHONE=de_DE.UTF-8
LC_MEASUREMENT=de_DE.UTF-8
LC_IDENTIFICATION=de_DE.UTF-8
LC_ALL=

Video Problems

To be honest, my first experiences with the Raspberry Pi went quite well but it was rather surprising that the only area that did cause problems was video.

It might be a peculiarity of the monitor that I am using but plugging in the video cable after the Raspberry Pi was booted up did not output any video. It was only with a restart of the Raspberry Pi with the HDMI cable connected was the video displayed on my monitor.

I have not had any problems connecting the Raspberry Pi to a device with a HDMI input only with slightly older equipment such as a monitor with only a VGA input. In this case nothing was displayed on the monitor at all. The solution is to modify the setup to a specific HDMI mode, which will essentially force VGA.

Of course, it is not intuitively obvious how to make this change if no output is being displayed on the monitor. The configuration for some of the system settings is the config.txt file which is in the /boot directory after booting up your Raspberry Pi. This is a catch 22 situation because if you video isn't working you will never be able to edit this file.

However, you can edit the file on the SD card on another computer that is running either windows or Linux. The SD card will have a partition named boot which will contain a couple of files that are used during the bootup. The only file that is of real interest to us is the config.txt file, which later will be in the /boot directory on our Raspberry Pi.

This file is only about 40 lines long and a small change is required on only two of them. The two lines are actually some video settings that are commented out be default, which can simply be changed to force the video to VGA.

Before
uncomment to force a specific HDMI mode (this will force VGA)
#hdmi_group=1
#hdmi_mode=1

After
# uncomment to force a specific HDMI mode (this will force VGA)
hdmi_group=1
hdmi_mode=1

It is a simple matter of opening up this file in an editor and removing the hash(#) symbol from in front of those two lines.

In Unix shell scripts, lines that begin with a hash(#) symbol are comment lines. In this manner it is easy to have configuration files defined with some optional lines pre-prepared with alternative configurations or values. This is the exact same situation we see with this configuration file, the options that exist are listed and can be modified to have a slightly different behavior.

This configuration file contains more than just video settings, it is near the bottom of this file where you can see the configuration that is used when over-clocking the Raspberry Pi.

Everything sounds so simple with a small list of todo's. However, some windows users might have noticed that the contents of this file look really odd with all the characters running together into a single long line.

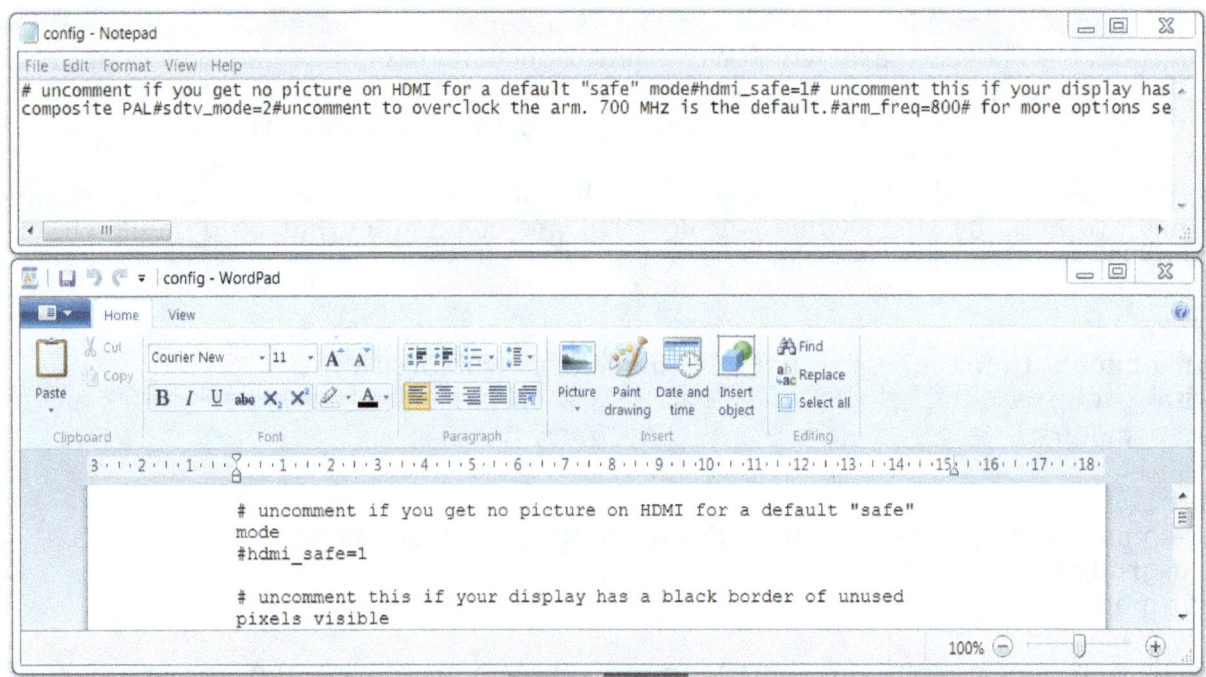

The reason for this is due to windows and Unix having different line endings[41], and windows default text editor notepad (notepad.exe) cannot properly edit these files. Notepad can only properly view and edit files with carriage return + line feeds. Microsoft does ship an editor which can edit both Unix or Windows files, it is called WordPad (write.exe).

If for some reason WordPad is not acceptable or more features are desired there are quite a few other text editors available for download from the Internet. One of the more popular editors that has no problems with either Windows or Linux file endings is Notepad++[42].

IP address Problems
I don't have an ip address or I don't have an internet connection. Well, if you run ifconfig from the command prompt and get output that is similar to that below you probably don't have an ip address.

```
$ ifconfig
eth0: flags=4163<UP,BROADCAST,RUNNING,MULTICAST>  mtu 1500
        inet6 2003:cb:173d:4900:990d:471:a9dd:cbe3  prefixlen 64  scopeid 0x0<global>
        inet6 fe80::cf0e:a3bd:d497:e8d4  prefixlen 64  scopeid 0x20<link>
        ether b8:27:eb:d5:c1:b7  txqueuelen 1000  (Ethernet)
        RX packets 3577  bytes 416220 (406.4 KiB)
        RX errors 0  dropped 23  overruns 0  frame 0
        TX packets 1131  bytes 240165 (234.5 KiB)
        TX errors 0  dropped 0 overruns 0  carrier 0  collisions 0

lo: flags=73<UP,LOOPBACK,RUNNING>  mtu 65536
        inet 127.0.0.1  netmask 255.0.0.0
        inet6 ::1  prefixlen 128  scopeid 0x10<host>
```

41 http://en.wikipedia.org/wiki/Newline

42 https://notepad-plus-plus.org/

> *loop txqueuelen 1000   (Local Loopback)*
> *RX packets 0   bytes 0 (0.0 B)*
> *RX errors 0   dropped 0   overruns 0   frame 0*
> *TX packets 0   bytes 0 (0.0 B)*
> *TX errors 0   dropped 0   overruns 0   carrier 0   collisions 0*

If when running this command you see a line that looks similar to the following then you do have an ip address.

inet 192.168.178.57   netmask 255.255.255.0   broadcast 192.168.178.255

There is one special exception, that is the address 127.0.0.1. This is the loopback address for your Raspberry Pi and should always exist.

If you do not have an ip address this means that you are not connected to any device with a DHCP server running or it might be that the Ethernet cable is not plugged in. A DHCP server is usually running on the server(work) or your router(home) to give out ip addresses to any machines that ask.

It seems like just about any and every networking product come with their own DHCP server installed, this may be a smart switch, Network Attached Storage device or your router. If you are not getting an ip address your problems are most likely not on the Raspberry Pi.

Hardware Attached Top
The Arduino is a great micro-controller but as it is only a micro-controller board it was common to extend it by adding a "shield". A shield is an addon-board which is connected to the Arduino on top of the header pins. The ability to change between different shields each of which can provide different functionality is a powerful feature.

The Raspberry Pi was originally created in 2012 and although it did have header pins to allow easy access to external peripherals no equivalent for the "shield" existed. This was due to the fact that the Raspberry Pi was not expected to become such a big hit. The Raspberry Pi was released at exactly the right time and the interest exploded.

The Raspberry Pi Foundation is a clever bunch with an eye on the future and they created the Hardware Attached Top (HAT) for the model B+.

The Raspberry Pi B+ also came out in 2012 which expanded the number of pins from 26 to 40. This Pi was built with the intention of supporting shield addon boards (a "Hat") similar to the Arduino shields. Increasing the number of pins was a minor upgrade but the number of pins also was used as part of the HAT functionality.

Prior to the HAT being introduced the Raspberry Pi itself had no knowledge if a board was connected or not. It is the responsibility of the user to install any necessary drivers or modify any configuration files. One of the main features of this new functionality is that the Pi can identify the HAT that is connected and install drivers and configure itself.

Raspberry Pi Zero
The history of the Raspberry Pi has been a journey of incremental improvements. The Foundation initially released two versions each with a different set of features. Subsequent releases increased the hardware specifications in terms of memory, cpu cores, and speed. But one of the main features of the Raspberry Pi has been compatibility.

With the release of the B+ the header pins did expand from 26 to 40 but even on the new B+ the original 26 pins kept the same functionality. Thus if a specific custom board was developed for an older Raspberry Pi it should still work on the newer versions. One other type of compatibility, although somewhat less important, is the physical size of the Raspberry Pi. All Raspberry Pi's up until the Raspberry Pi Zero was the same size PCB. This made it easier to reuse old cases(85mm x 56mm), boards and simply upgrade your project with new hardware without requiring any other changes.

The Raspberry Pi Zero, shortly followed by the Raspberry Pi Zero W was a brilliant addition to the Pi family. The new size of the board is 65mm x 30mm. This is almost a third of the size of the original pi but it contains all the horsepower of the latest generation Raspberry Pi 3. It does have slightly less memory and only one micro-USB but it does support both Bluetooth and WiFi.

The Raspberry Pi Zero W also has 40 pins and is also compatible with the hardware attached top. Regular HAT's are just a bit too large which is why there also exists a similar pHAT which is the same as a regular HAT but it is sized for a Raspberry Pi Zero. These smaller pHAT's have been also referred to as bonnets on the Adafruit site.

Troubleshooting Video problems
If you can setup one Raspberry Pi you can usually setup any Raspberry Pi however there is one other small difference between the Zero and the standard pi. The standard pi also has a couple of LED's which flash during the boot process as well as showing Ethernet activity.

It can be harder to diagnose when difficulties arise because of this limited feedback. The Raspberry Pi zero, unlike the other models actually requires that video drivers are loaded before output will be displayed on the monitor. If for some reason you are having difficulties reading / booting the operating system it is possible you will not see anything on the screen at all.

Summary
The Raspberry Pi is a great solution for low cost projects that need more resources than can be offered by a Arduino. One of the biggest advantages is that the Raspberry Pi has access to an extremely large body of open source software. It can act as a node on a network and interact with other computers using many popular communication protocols (WiFi, Ethernet, Bluetooth) and in many cases accept any type of USB based devices.

The Raspberry Pi isn't a calculator that can also run tiny programs but a full fledged computer that can run a web server or act as a stand alone desktop to be used for smallish development projects.

8 STREAMING AUDIO

I was at work when I was hit with the following wordquake.

"Is that a mp3 player? I haven't seen one of them in a long time"

It was lucky that my coworker wasn't a recent graduate but rather a mature older fella. Yet it did get me to thinking of how things have changed in even the last few years. Companies that rented dvd's have been going out of business, some stores have reduced or eliminated the sales of music cd's and not streaming movies is now considered a bit odd.

That got me to thinking about the humble radio. I used to listen to it while falling asleep but it would be nice if it shutdown after I fall asleep. The old solution to this problem might have been to plug the radio into a timer.

A newer high tech solution might be instead to stream the music. The internet is full of every type of music imaginable, not only would I have access to a wide variety but the device could be programmed to stop after an hour. There is no existing program that I am aware of that will do this for me but with the power of open source software and with a few creative this can be solved. Not only that it can be done in a fraction of the time it would take to write some monolithic program to do this.

The Raspberry Pi is a tiny computer that runs a fledged version of Linux. Nobody buys a computer because of the operating system but rather what software it can run. The software repository for the Raspi is very comprehensive and provides all the building blocks necessary for this project.

1. GUI for Music
2. Program to stream music

GUI
The actual graphical part doesn't need to provide all that much in functionality. It simply needs to allow me to select which station I want and then to act on that. I am not creating a full fledged internet radio so I will skip right to the part where I develop a GUI as essentially a bunch of station presets.

Music selection

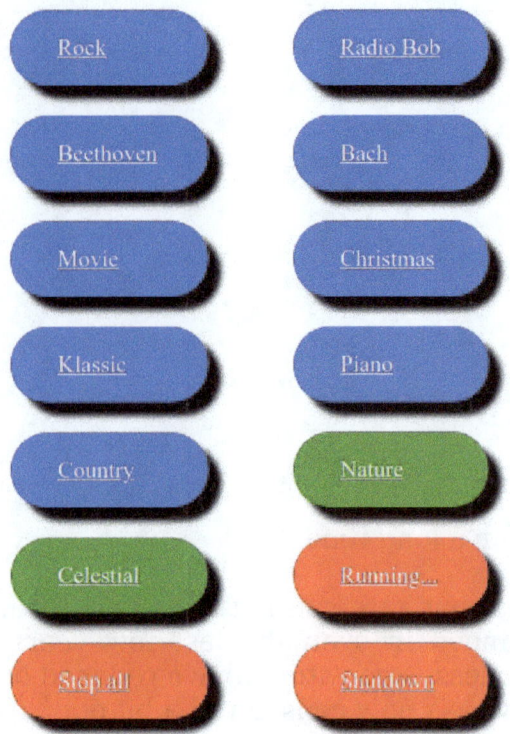

This can easily be done using HTML and can be delivered to the user via the Apache web server. This is a pretty neat solution as it can be viewed by not only desktops and laptops but also by tablets and smart phones as well if they are on the network.

Audio Player
The requirements for a music player are both very high and very low. Obviously if we are to stream music from the internet the player will have to be able to accept URL's as the input. Oddly enough no other GUI requirements are necessary. The station preset is controlled by the web browser so no GUI at all is required, a simple command line program which plays the music streams would not only be good it would be perfect.

The Raspi software repository has a couple of different players available.

    omxplayer
    mplayer
    aplay
    mpg321

I have used the mplayer in the past so I will continue to use that for this project. You simply run the program with the URL as the input file.

    e.g.
        mplayer http://bob.hoerradar.de/radiobob-live-mp3-hq

This will play music from this source. One of the other functions of this player is that it can also play normal mp3 files as well. This offers the flexibility that in addition to streaming music it can also play audio files. This might be nice if you want to listen to some form of white noise or nature sounds.

The only problem is that although it is possible to run scripts from Apache, it is not possible to run the mplayer from one of those scripts. This is not actually due to a weakness of Apache but rather a strength. The default security setup has been done in such a way that the process that runs the script has very few privileges. One solution would be to grant a lot of unnecessary privileges to that user (www-data) but a better solution would be to save the requests and then run them from another user that has the correct permissions.

Scheduling software

There is no current open source project that would fulfill this particular request. The task is to essentially create a list of music or streaming URL's to play. Processing this list means to take and remove the first item from the list while new requests would be added to the bottom of the list. This description is identical the abstract data type first in first out (FIFO) queue which is taught to all beginning computer science students.

The actual use of the FIFO queue would be broken up into two separate pieces, the first part is adding the new requests to the queue. This task will be executed by the web site script when the user presses one of the preset buttons. The value of the preset is actually the name of a configuration file which will contain the URL for that particular music channel.

The second half of the task is to remove items from the queue and then play that music. This actually would be somewhat counter intuitive if the only task of the player is stream music because the music stream in theory never ends. These presets could be URL's for a streaming station or they could be a mp3. If they are a mp3 they would finish and so the ability to have a queue full of items is an additional feature.

However, because of how the scheduler has been designed there have been a few additional buttons for adding control.

> Running
> Stop
> Shutdown

These "control" buttons perform the actions that reflected in their description. The "running" option will display the information about the currently running stream while the stop option will stop the currently running stream.

In order to realize my replacement radio solution I will have to implement the following steps.

1. Setup the Raspberry Pi[43]
2. Install Apache
3. Install mplayer
4. Enable CGI functionality for Apache
5. Create Web page

Install software

Apache web server and the music player are available in the repository. These can be installed on the Raspberry Pi with the following command.

---

43 How to setup a Raspberry Pi was discussed in chapter 7.

    sudo apt-get install apache2 mplayer

The installation of these components can be installed separately but it isn't necessary.

Configure Apache CGI

CGI stands for Common Gateway Interface which is a way for the web server can interact with other programs for generating content. These other programs are often referred to as CGI scripts. These scripts make it easy to incorporate external content to the web site such as data from a database. It is possible to configure Apache use multiple file extensions to allow these scripts to be written in a programming language you are most comfortable with. Apache actually supports multiple modules, CGI is just one of them.

The web server is installed to the /etc/apache2 directory. In this directory are a few other important sub-directories.

    conf-available
    conf-enabled
    mods-available
    mods-enabled

When Apache is installed it also installs a few scripts which can be used for the configuration of Apache. The a2enmod can be used to enable the cgi command.

    sudo a2enmod cgi

However, once this module has been enabled one small change to the configuration files is needed. The file serve-cgi-bin.conf is in the /etc/apache2/conf-enabled directory.

Before
```
<IfModule mod_alias.c>
    <IfModule mod_cgi.c>
        Define ENABLE_USR_LIB_CGI_BIN
    </IfModule>

    <IfModule mod_cgid.c>
        Define ENABLE_USR_LIB_CGI_BIN
    </IfModule>

    <IfDefine ENABLE_USR_LIB_CGI_BIN>
        ScriptAlias /cgi-bin/ /usr/lib/cgi-bin/
        <Directory "/usr/lib/cgi-bin">
            AllowOverride None
            Options +ExecCGI -MultiViews +SymLinksIfOwnerMatch
            Require all granted
        </Directory>
    </IfDefine>
</IfModule>

# vim: syntax=apache ts=4 sw=4 sts=4 sr noet
```

After
```
<IfModule mod_alias.c>
    <IfModule mod_cgi.c>
        Define ENABLE_USR_LIB_CGI_BIN
    </IfModule>

    <IfModule mod_cgid.c>
        Define ENABLE_USR_LIB_CGI_BIN
    </IfModule>
```

```
        <IfDefine ENABLE_USR_LIB_CGI_BIN>
            ScriptAlias /cgi-bin/ /usr/lib/cgi-bin/
            <Directory "/usr/lib/cgi-bin">
                AllowOverride None
                Options +ExecCGI -MultiViews +SymLinksIfOwnerMatch
                Require all granted
                Order allow,deny
                Allow from all
            </Directory>
        </IfDefine>
</IfModule>

# vim: syntax=apache ts=4 sw=4 sts=4 sr noet
```

Once this is done it is possible to put scripts into the /usr/lib/cgi-bin which can then be called.

When modifying the configuration files you need to restart Apache.

sudo systemctl restart apache2

Create a web page

The web page is important as this is the visual piece that the user will interact with however the Raspberry Pi is a lightweight piece of hardware. The GUI is only used for launching the music so the solution should not be written in such a way to overly tax the hardware.

I have written the GUI using just a small bit of HTML and a style sheet.

Index.html

```
<html>

<head>
<title>Boys room</title>
<link rel="stylesheet" href="style.css">
</head>

<body>

<h1>Music selection</h1>

<!-- table starts here -->

<table class="w3-table-all" style="width: 439px; height: 177px;">
<tbody>

<tr>
  <td>
    <div class="button-wrapper">
    <a class="button music-button" href="cgi-bin/scheduler.py?name=rock.txt">Rock</a>
    </div>  
  </td>

  <td>
    <div class="button-wrapper">
    <a class="button music-button" href="cgi-bin/scheduler.py?name=radiobob.txt">Radio Bob</a>
    </div>  
  </td>
</tr>


<tr>
  <td>
    <div class="button-wrapper">
    <a class="button music-button" href="cgi-bin/scheduler.py?name=beethoven.txt">Beethoven</a>
```

```html
        </div>  
      </td>

   <td>
      <div class="button-wrapper">
        <a class="button music-button" href="cgi-bin/scheduler.py?name=bach.txt">Bach</a>
        </div>  
   </td>
</tr>

<tr>
   <td>
      <div class="button-wrapper">
        <a class="button music-button" href="cgi-bin/scheduler.py?name=movie.txt">Movie</a>
        </div>  
   </td>

   <td>
      <div class="button-wrapper">
        <a class="button music-button" href="cgi-bin/scheduler.py?name=christmas.txt">Christmas</a>
        </div>  
   </td>
</tr>

<tr>
   <td>
      <div class="button-wrapper">
        <a class="button music-button" href="cgi-bin/scheduler.py?name=klassic.txt">Klassic</a>
        </div>  
   </td>

   <td>
      <div class="button-wrapper">
        <a class="button music-button" href="cgi-bin/scheduler.py?name=piano.txt">Piano</a>
        </div>  
   </td>
</tr>

<tr>
   <td>
      <div class="button-wrapper">
        <a class="button music-button" href="cgi-bin/scheduler.py?name=country.txt">Country</a>
        </div>  
   </td>

   <td>
      <div class="button-wrapper">
        <a class="button whitenoise-button" href="cgi-bin/scheduler.py?name=nature.txt">Nature</a>
        </div>  
   </td>

</tr>

<tr>
   <td>
      <div class="button-wrapper">
        <a class="button whitenoise-button" href="cgi-bin/scheduler.py?name=celestial.txt">Celestial</a>
        </div>  
   </td>

   <td>
      <div class="button-wrapper">
        <a class="button control-button" href="cgi-bin/cmd_running.py">Running...</a>
        </div>  
   </td>

</tr>


<tr>
```

```html
        <td>
          <div class="button-wrapper">
            <a class="button control-button" href="cgi-bin/scheduler.py?name=cmd_stop.py">Stop all</a>
          </div>  
        </td>
        <td>
          <div class="button-wrapper">
            <a class="button control-button" href="cgi-bin/shutdown.pl">Shutdown</a>
          </div>  
        </td>
      </tr>

    </tbody>
  </table>
  </body>
</html>
```

This style sheet is just a convenient way to define a couple of colored buttons. Each of these colors are intended to represent a different type of functionality.

Blue	music stream
Green	music file
Red	system buttons

style.css

```css
# received hints from
# http://stackoverflow.com/questions/27943221/bootstrap-background-color-changes-on-active
p {
  color: red;
}

a {
  color: blue;
}

.button-wrapper {
  display: block;
  text-align: left;
}

.button {
  border: none;
  border-radius: 3em;
  box-shadow: 10px 10px 5px black;
  display: inline-block;
  font-size: 18px;
  padding: 1em 2em;
  width: 80px;
}

.button:hover {
margin: 5px auto 3px;
-moz-box-shadow: 0 0 1px #eee;
-webkit-box-shadow: 0 0 1px #eee;
-o-box-shadow: 0 0 1px #eee;
-ms-box-shadow: 0 0 1px #eee;
box-shadow: 0 0 1px #eee;
}

.control-button {
  background-color: red;
  color: #fff !important;
}

.music-button {
  background-color: blue;
  color: #fff !important;
```

```css
}

.whitenoise-button {
  background-color: green;
  color: #fff !important;
}

.cta-button:hover {
  background-color: red;
}
```

Create a scheduler

The scheduler for this particular process is very simple. The process is nothing more than creating a list of tasks and periodically taking a task from the list and executing it. Because this application is intended to be similar to a radio the number of items in the scheduling list would normally be one, but perhaps it could be as many as two if the user changed his mind and then scheduled the stop all process.

Adding a value to the queue really shouldn't require more than a few lines.

scheduler.py

```python
#!/usr/bin/env python

import sys
import cgi
import cgitb; cgitb.enable() # Optional; for debugging only
import datetime
import time


playerwd="/piplayer";
queuename=playerwd + "/cache/piplayer.queue.txt";
runtime=playerwd +  "/cache/runtime.txt";


def addtoqueue(nametoadd):

        fh = open(queuename,'a')
        fh.write('{0:s}\n'.format(nametoadd))
        fh.close()
        return

def main():
        debug=0

        print "Content-Type: text/html" # html content to follow
        print                   # blank line, end of headers

        if (debug == 1):
                musictype = "test_name.txt"
        else:
                musictype = "willbeoverwritten.txt"
                arguments = cgi.FieldStorage()
                musictype = arguments['name'].value;

        print "<h1> scheduling " + musictype + "</h1>";
        addtoqueue(musictype)

main()
```

The process of queuing a new command is simple but the task of processing those commands is a bit more complex.

taskmgr.py

```python
#!/usr/bin/python
```

```python
import sys;
import os;
import time;

#name of our audio player
musicplayer="vlc";
musicplayer="mplayer";

playerwd="/piplayer";
playerconfig=playerwd + "/config/";
playerclient=playerwd + "/client/";
runtime=playerwd +  "/cache/runtime.txt";

#input queue
queuename=   playerwd + "/cache/piplayer.queue.txt";

#queue less the top item
queuenametmp=playerwd + "/cache/piplayer.queue.tmp";

def playerrunning(player):
        retval=0
        cmd="ps -ef | grep " + player + " | grep -v grep > /tmp/running.txt"
        os.system(cmd);

        fh = open("/tmp/running.txt","r")
        contents =fh.read()
        fh.close()

        if (len(contents)):
                retval=1

        return retval;

def pop_top():
        # get top item from queue and write
        # the rest into temp file, then
        # move temp file to queue

        # debug
        # print queuename;

        fhi = open(queuename,"r")
        fho = open(queuenametmp,"w")
        tinputmp3 = fhi.readline()
        tinputmp3 = tinputmp3.rstrip();

        for x in fhi:
           fho.write("{0:s}".format(x))
        fho.close()
        fhi.close()

        cmd="mv " + queuenametmp + " " + queuename
        # print cmd;
        os.system(cmd);
        return tinputmp3;

def dotask():
        # if there is an item in our queue
        #
        # now that we have our file, move the
        # remaining queue items to the queue
        # from the working file
        #

        inputmp3 = pop_top();
        # print "input3 = " + inputmp3;

        if (len(inputmp3)>0):
           # debug
```

```
            # print inputmp3

            cmdcheck=inputmp3[0:4]

            if (cmdcheck == "cmd_"):
               # print "this will be a command"
                    cmd=playerclient + inputmp3
                    # print cmd
                    os.system(cmd);
            else:
                    # if something is playing ... then stop it
                    if (playerrunning(musicplayer)>0):
                            cmd=playerclient + "cmd_stop.py"
                            # print cmd
                            os.system(cmd);
                            time.sleep(3);

                    #
                    # get actual contents of what to play
                    #
                    fhurl = open(playerconfig + inputmp3,"r")
                    actualurl = fhurl.readline()
                    actualurl = actualurl.rstrip();
                    fhurl.close()
            #
            # now we need to run the actual item.
            #
                    cmd = "nohup " + musicplayer + " " + actualurl + ">/dev/null 2>/dev/null" + "&";
                    print "playing ... ";
                    os.system(cmd);
            else:
                    print "nothing to play"
            return ;

      dotask();
```

Part of the reason that this is more complicated is because the tasks are mixed. The are commands to be executed or actual data to be passed to the music player. To simplify this task, the commands all begin with a special prefix in order for the script to easily separate the two types of data.

Another complication is that you cannot play a second source of music while the first one is playing. This is not necessarily a limitation of Linux but has more to do with the fact that playing two different radio stations at the same time would be irritating not relaxing.

The task scheduler will execute a command or play the music each time it is run but it needs to be run in order to do so. Linux has a scheduler, called cron, which will launch programs on a scheduled basis. I simply add the taskmgr.py to the crontab – the crontab is the name of the list of tasks to be run by cron.

```
Crontab value
# m h  dom mon dow   command
* * * * * su -c '(umask 002; /piplayer/client/taskmgr.py)' - pi
```

Music stations

One of the points not elaborated on to this point is the actual source of music. The webpage, index.html, is not only just a pretty GUI nor does it contain hard coded values for the music. The webpage actually points to a configuration file that contains the source for the music for each button.

This source can be anything that can be played by the mplayer application. This is actually very convenient as this can be either a local path to the music files or URL's to a streaming music source.

>bach
>https://klassikr.streamabc.net/klr-purebach-mp3-128-7912278
>
>celestial.txt
>/piplayer/files/Celestial.mp3

This actually sounds pretty simple and for local music files this is extremely simple but what about streaming music sources? The first time I worked on this project I could actually find a few websites that were very open with their streaming URL's but the second time around this information was harder to acquire.

I did manage to discover a easy way to locate the URL. Simply use your web browser to navigate to a site and play one of the streams from that site. Once the streaming has started in your browser then open the developers tool by pressing the F12 key. This opens up the developers tools that allows the developer to examine virtually all aspects of the browsing process. This can be the HTML or CSS code or it can be which types of HTTP calls are being done along with the response status code for this call.

This also allows you to see what network URL's have been called. This makes it trivial to retrieve and then use the URL as a pre-configured value for one of the configuration files.

Complications

In general these scripts are not so complicated but it is the various operating system permissions or lack thereof that made all the problems.

When Apache is configured to allow scripts to be run it creates the opportunity for a poorly written script to gain elevated system privileges. This is a known concern and so when this functionality was added to Apache it was designed that these scripts would be run as a different user, www-data, instead of the user running the Apache service.

The www-data user has been configured so he cannot even log into the computer. This user cannot actually run the music player. The work around for this is for a normal user to run the music player. The good news is that adding values to our list is simply writing values to our "queue". No particular special permissions are required to do that.

Beyond that, there were some minor complications with getting the permissions and user groups setup correctly. The "client" portion of this audio solution is running as the "pi" user which is not a privileged user.

Because of the fiddly nature of this setup, I had to create a script in order to make sure that it was always done in the exact same/correct manner.

Install script

```bash
#!/bin/bash

CGIBIN=/usr/lib/cgi-bin

mkdir -p /piplayer/cache
mkdir -p /piplayer/files
mkdir -p /piplayer/client
mkdir -p /piplayer/config
```

```
chown -R www-data /piplayer
chgrp -R www-data /piplayer
chmod -R 755 /piplayer

chown -R pi /piplayer/cache
chmod -R 770 /piplayer/cache

usermod -a -G www-data pi

cp files/* /piplayer/files
cp client/* /piplayer/client
cp config/* /piplayer/config
cp cgi-bin/scheduler.py $CGIBIN
cp cgi-bin/cmd_running.py $CGIBIN
cp html/* /var/www/html

touch /piplayer/cache/piplayer.queue.txt
chown www-data:www-data /piplayer/cache/piplayer.queue.txt
chmod 777  /piplayer/cache/piplayer.queue.txt
```

9 NETWORK STORAGE

There are several large open source project that have such a high visibility that almost everyone has heard about them. Here is a list of some of these projects in no particular order.

    Linux
    Gimp
    Firefox
    Open Office
    Eclipse
    VLC
    XBMC / Kodi
    Open Stack
    Apache

This is not an extensive list but most people have heard of at least half of them. There is another project that might not have this same level of visibility with the average person but is both known and probably used in almost all fortune 500 company computer centers. This project is SAMBA.

Samba is the implementation of the Server Message Block (SMB) application layer[44] networking protocol. Perhaps easier to understand would be to think of this as a file sharing protocol. This enables client computers the ability to read or write files to a remote computer. This particular protocol was initially written in the 1980's and also supported viewing and printing to registered printers.

This protocol was enhanced over the years from SMB to SMB v2 and SMB v3.

SMB	SMB v2	SMB v3
Maximum block size 64K	Introduced with Windows Vista	Introduced with Windows 8
UNI code support (added later)	Durable file handles	SMB Direct protocol
shared access to printers and serial ports	Supports symbolic links	SMB Multichannel
	Remove block size limitations	SMB Transport failover

---

44 The application layer is the upper most layer in the 7 layer OSI networking model. This topic, although interesting, is not one that most people give any consideration to. The OSI model shows the different abstraction layers that data will go through as it traverses from the user or program down to the physical hardware and over the network to another user or program.

Better locking mechanisms   AES 128 CCM encryption

SMB v3.1.1
Introduced with Windows 10 /
Windows server 2016
AES 128 GCM encryption
integrity check using SHA-512 hash

What makes Samba interesting is that is allows better communication between computers especially between Linux and Windows. Over the decades Samba has gone from the original free software implementation of the SMB protocol by Andrew Tridgell into what can only be described a fully compatible family member to the Windows operating system dynasty.

The current version of Samba at this writing is Samba 4 which supports all four of these protocol versions.

>   NetBIOS
>   NTLM
>   Security Account Manager database
>   WINS
>   Distributed File System
>   Active Directory Domain Controller

What this list means is that it would be possible to have the functionality of a Windows network server but without the necessity of purchase Windows server. This is true even if your company is powered by a computer or operating system that is not associated with Microsoft. Samba 4 runs on Unix, Linux, OpenVMS, Solaris, AIX, macOS Server and BSD variants.

Samba on Raspberry Pi
The Raspberry Pi has quite a few different distributions of Linux that can be installed on it for general purpose computing but it also has a few special versions intended to support retro gaming or to support your own media center.

>   Raspbian (modified Debian)
>   Ubuntu Mate
>   Snappy Ubuntu Core
>   PINet
>   Windows 10 IOT core
>   RISC OS
>   OSMC
>   Recalbox
>   Lakka
>   libreElec
>   TLXOS

Most of these operating systems are geared to a specific purpose but it is possible to simply install Raspbian or Ubuntu Mate in order to have your own general purpose computer. With that as a base it would be completely possible to install Samba 4 on the computer and create your own active directory domain controller for your network. Although this is a possibility, I am not going to create a full fledged Samba powered domain controller.

Is it because the Raspberry Pi isn't up for the task? Actually this is not the case. It was only a few years ago when a client of mine was adding an additional drive to their companies file server. If memory serves, the computer was a 1,4 ghz server with 4 gb of ram.

This is not very far off the basic requirements for a computer running windows 2008 server.

    1 - 2 Ghz x86 processor
    512 MB - 2 gb ram
    100 Megabit Ethernet

The Raspberry Pi 3+ has similar but not quite the same specifications.

    1.4 GHz 64bit processor[45]
    1 GB ram
    1 Gigabit Ethernet

The Raspberry Pi does not have terabyte disk drives and to be honest the throughput in heavy traffic situations would be somewhat lacking, but it should be possible to create an active directory domain controller[46] which could easily deal with the logins. Not only that it would be easily possible to also have a small pile of Raspberry Pi's as backup domain controllers.

Most people don't need a domain controller at home and quite a few home computers actually are running with home editions of Windows. Windows home editions has been sold at a reduced cost to people who don't really need this functionality but to keep their other customers from using this cheaper version of windows they have turned off[47] this domain functionality. This actually prevents those computers from joining such a domain as you would probably see in an office.

    https://www.kerstner.at/2018/11/setting-up-an-active-directory-domain-controller-with-samba-4-on-a-raspberry-pi-3/

My example will be less complicated but perhaps more applicable for more people.

File server
We can implement a subset of the full Samba functionality. It is possible to create a file share that can be mounted on either a Linux or Windows pc.

The Samba project actually does all the hard work for us. There are only a few configuration steps needed.

    Samba server
        1   Install Samba
        2   Setup configurations

---

45 Cortex-A53 (ARMv8) 64-bit SoC @ 1.4GHz

46 https://www.kerstner.at/2018/11/setting-up-an-active-directory-domain-controller-with-samba-4-on-a-raspberry-pi-3/

47 Crippled in my opinion.

3. Add a samba user
4. Restart samba

It may seem a bit counter intuitive but there is no package for windows. Samba is only available for Unix like operating systems plus OpenVMS and macOS because windows server already supports the SMB protocol.

It shouldn't be necessary to download a Samba package as Samba will most likely be already in the package repository of your Linux distribution. On a Debian based system installing this is as matter of the following command.

    sudo apt-get install samba samba-common-bin

The Samba service has a configuration file named smb.conf which is located in the /etc/samba directory. This configuration file can be setup to create roaming profiles or file shares. There have been many books on this topic with each being hundreds of pages.

Considering the level of functionality available it is not possible for me to fully explain all of the various configurations in this book but I can explain the necessary setup for creating a simple Samba share.

Each share has its own section much like a Windows INI file. Each section has a section name in square brackets. This portion of the smb.conf file is for a Samba share that is not so creatively named "share".

In Table 18 Samba share configuration is the extracted configuration from a smb.conf file for a Samba share called "share" with each line explained.

Configuration	Description
[share]	Name of the samba share
comment = Pi shared folder	Description about samba share
path = /share	This is the path on the server to where the samba data is stored.
browseable = yes	When browseable is equal no the user must already know the name of the share, e.g it is not discoverable.
writeable = yes	Not readable
only guest = no	Also samba users
create mask = 0777	This is the maximum permission granted to a newly created directories
directory mask = 0777	This is the maximum permission granted to a newly created directories
public = yes	When set to No requires a username and password to access samba share
guest ok = yes	Yes means that it is possible for a user to write to the samba share without actually having a samba user.

Table 18: *Samba share configuration*

Adding a user to the Samba database of users is very reminiscent of changing a regular Linux users password.

    $ sudo smbpasswd -a pi
    New SMB password:
    Retype new SMB password:
    Added user pi.

$

This example was creating the user pi, with the password samba. Just like changing normal Linux passwords the password entered is not echoed to the screen.

Once the samba configuration has been finished, either the samba daemon needs to be restarted or the computer would need to be restarted. The command for a Raspberry Pi is as follows.

    sudo /etc/init.d/samba restart

This pretty much concludes the setup of a Samba share on a Linux machine. Once the restart is finished you can begin to use this new Samba share on your windows computers.

Linux Client
Creating a network share is not uncommon in the Unix or Linux world however creating a samba share is usually done with the intention of sharing files between Linux and Windows computers. When Windows is taking out of the equation then sharing files between Unix like systems would probably use a Network File System (NFS) not Samba but ....

There is yet another protocol that does help make Samba convenient even on a Linux machine it is common internet file system or CIFS, which happens to be a dialect of the SMB protocol. These protocols are close enough that they can be used interchangeably. This is important as CIFS shares can be mounted directly as a drive to a Linux mount point.

Just like any normal Linux drive there are a few steps that need to be performed prior to mounting a disk.

    1   Create a mount point
    2   Configure drive mounting

Mount Point
Under windows, a new disk drive is simple given an unused drive letter. Unix like operating systems have a much slicker approach. You simple create an empty directory with the name you want and then attach or mount a drive to that subdirectory. Once a disk has been mounted any files created or deleted in that directory will actually be created or deleted on the disk but the file system feels like a single complete drive. This empty directory is called a mount point.

    create a mount directory
    sudo mkdir -m 1777 /share

Creating the directory in this manner, unlike the permissions described in the appendix, creates a directory with the additional attribute of a sticky bit. This bit means that files or directories can only be renamed or removed by the owner of the file, directory owner or root user.

One other special permission that can be used when creating directories would be the setgid capability. This permission causes all new files to be created with the same group as the parent.

    sudo mkdir -m 2777 /share

Once everything is prepared then mounting a disk is done using the mount command.

    sudo mount /dev/sda1 /media/mymountpoint

This will assign the device /dev/sda1 to the path /media/mymountpoint. In the best case, no additional parameters are needed. The mount command will mount a device but when no parameters are given you will see a list of all mounted devices on the system. Below is the sample output from the mount command for the /dev/sda1 disk.

    /dev/sda1 on /media/mymountpoint type vfat (rw,relatime,fmask=0022, dmask=0022,codepage=43,iocharset=ascii,shortname=mixed,errors=remount-ro)

In this case the device mounted was a USB stick and we can see from the mounting information it was a windows file system. Looking at the mounted device we can see that the file system is vFAT which is an extension to the Windows 95 FAT file system but with 255 byte long filenames.

The rest of the values affect how the read and write actions will be treated. The relatime option will prevent the access time from being written to the filesystem during every the file accessed. When mounting a drive it is better to specify what behavior is desired rather than let the operating system default these values for you.

Normal users do not have the necessary privileges for simply mounting an unknown filesystem and so cannot mount any drives that have not previously been configured to for mounting. There are a few exceptions to this such as cd or dvd drives which can automatically mount themselves. The configuration of drives that can be mounted by users without enhanced privileges is done by adding the drive to the fstab file in the /etc directory. This file contains a list of which drives should be mounted, where they should be mounted and with what parameters.

    fstab example with device
    <file system>   <dir>   <file system type>   <options>   <dump> <pass>
    /dev/sda1   /media/mymountpoint   ext4   defaults   0   2

Once a device is in the fstab file then it is possible for a normal user to mount the device simply by using the mount command and the device name. The mount program will get the rest of the mounting values from the fstab file.

This information about mounting is important as it is also possible to mount a samba drive as a normal disk.

    fstab example with samba share
    //192.168.178.63/share   /mediamymountpoint   cifs   gid=100,uid=1000,user=pi,pass=samba 0 0

This example is actually not any different than mounting a device as a drive. The device is simply replaced with the share from the machine hosting it. The file system is of type "cifs". In this case the options are specifying the file system is mounted as user (uid = 1000) and group (gid=100) as well as the samba credentials(user=pi, pass=samba) necessary for accessing this share.

The credentials in this example is the user and password that was defined on the server with the smbpasswd command.

Special permission problems
Normally the file permissions on the mounted drive are used when it is mounted. If you mount a windows file system (type = vfat) there are some issues with permissions. The permissions for this mounted disk cannot be changed from rwxr-xr-x. This isn't so much a problem for the owner, in this case the owner is uid=1000 but more of a problem for users who are not in group gid=100.

The Raspberry Pi can be used as a cheap way to add additional disk space to your network as a primitive form of network attached storage. Yet this is not ideal for actually adding disk space in a production environment. The reason is not that this technology is not mature enough but that the Raspberry Pi doesn't have the same level of input/output throughput as a commercial solution.

Using a Raspberry Pi in this manner might be the perfect way to capture information on your network or enhancing an existing project but in either case it does extend the resources that your projects can utilize.

More information about file system permissions can be found in the section titled "Appendix - Linux Commands".

10 BUILD LED ARDUINO CUBE

Most electronics are fairly two dimensional. The PCB board and the components that are mounted on it all lie in the same plane which makes it essentially a two dimensional solution.

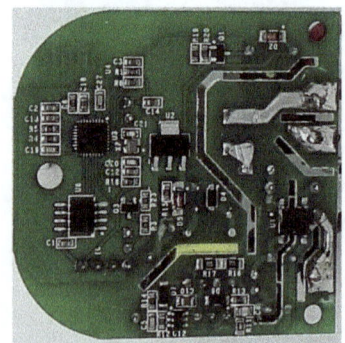

Sonoff S20 Board

Calliope micro-controller

555 project

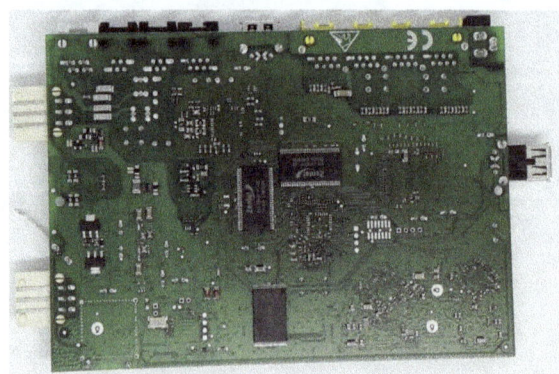

Fritzbox mainboard (backside)

Fritzbox mainboard (front side)

One interesting and fun electronics solution is a 3D LED cube.

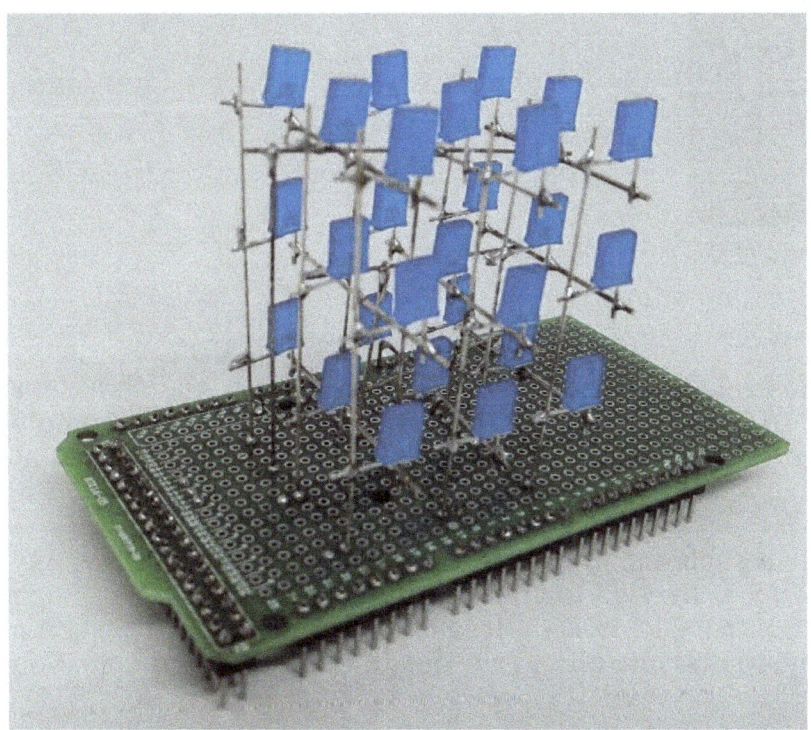
Illustration 94: Completed 3x3x3 cube Arduino shield

This has both electronics for the control but also the challenge of building a three dimensional LED cube that actually looks good. Typically a "good looking" cube is one that is fairly symmetrical.

Theory
The electronics behind a LED cube is actually not very complicated. The only tasks that it needs to do is to turn a LED on or off. When you are turning on and off a single LED this process is actually quite simple as there is only a single anode and a cathode but as the number of LEDs increases a clever design is necessary to control a single LED inside a large matrix of leads.

Conceptually the design for a LED cube layer is much like a checker board. You can control a much larger number of LEDs with a fewer number of leads when it is constructed in a grid pattern.

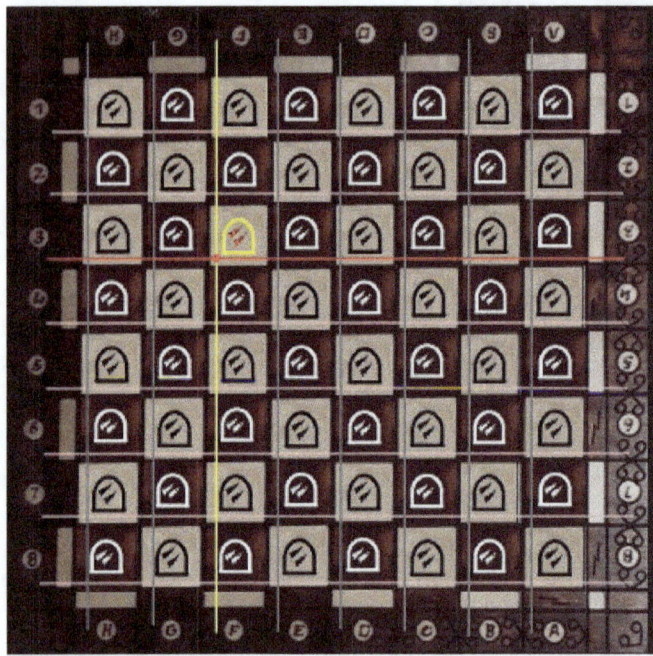
Illustration 95: Controlling a single LED from a LED grid

With only sixteen leads it would be possible to turn on or off any single LED in a sixty four LED grid. Simply supplying power to any of the leads columns (e.g A-H) while simply connecting leads for rows (e.g 1-8) to ground would enable the control of a single LED at the intersection.

Although 64 LEDs is a lot of LEDs for a such a small number of control wires this is a tiny number of LEDs when compared to what is needed to control a full 8x8x8 LED cube. Such a cube would have 1024 LED legs to control 512 LEDs. It is not possible to control such a large number of legs directly from any micro-controller. In order to control such a cube you would need a very clever design.

One such clever design for controlling an 8x8x8 cube is using a logic that is similar to that of the grid as seen in Illustration 95 but in a three dimensional way. That is to say that in addition to controlling which LED out of a single grid of 64 LEDs is lit, you would also control which of the eight layers would be lit at the same time. Simply using this exact logic would result in the requirement of 136 pins on your micro-controller.

```
16 leads in a layer  x  8 layers    = 128
1 lead   x 8 layers                 = 8
                                    = 136
```

This solution despite being much simpler than 1024 leads would still require a lot of wires carefully soldered together without any of them accidentally touching. This would also need to be done neatly so the cube looked good.

A further slight enhancement to that design could not only reduce the required pins but would also simplify the construction as well.

The simplification is to combine the ground pins from each level together and the entire level would be enabled or disabled. Doing this would mean that 64 columns are required and the anodes from each level are connected to them. The cathodes from each level would be connected together and also connected to the ground level control. Thus providing power to a given column and setting one or more levels to ground would allow the LEDs to be lit up while all other LEDs are not lit.

This setup allows you to turn on any single LED in the entire 8x8x8 cube with only 72 leads. Seventy two may seem like a lot of leads but this is actually much more manageable than 1024 or 136. This relatively small number of leads could be controlled by adding additional chips such as an I/O expansion controller.

LED cubes and patterns
LED cubes create neat images by turning on LED's in rapid succession in order to creates letters or patterns. The power required to turn on all the lights becomes more and more considerable as the cubes get larger and larger. A 3x3x3 cube can easily be powered by a 9 volt battery but as the number of LEDs increases (e.g. 8x8x8 or 16x16x16 cube) eventually we would require an actual power supply.

There are exceptions but essentially all cubes no matter their size behave using the same principles. That is each cube will generate their patterns by turning on a single LED at a time and rely on persistence of vision to fool the eye into believing that all the LED's in the pattern are on at the same time.

Persistence of vision principle is essentially the same concept underpinning "moving pictures" aka cartoons. The eye cannot distinguish between changes when they occur too quickly and so it will essentially blend them together to form a single picture. Thus when the different LED's in a cube switch rapidly on and off the eye simply sees the LED's were on.

One common example of this can be seen when using sparklers. When the sparkler is lit it can be moved through the air quickly and it will appear to the observer that the sparkler is leaving a trail in the air behind it.

Persistence of vision is important when creating more complex patterns. Some simple patterns can be done by simply lighting up all LEDs on a given level while others require that the different LEDs in the pattern are turned on and off repeatedly to generate the desired effect.

Simple 3x3x3 cube
Using an Arduino as the brains for the cube solves a lot of problems. It is a microcontroller platform that has its own integrated development environment and a rich ecosystem of libraries for many common tasks.

One of the tasks that is available "out of the box" from the Arduino is the ability to enable a pin for input or output and to send power (e.g turn it on). The next question is what should the circuit look like.

It is obvious that we need a small pile of LEDs and an Arduino but we will need a few other parts as well.

Bill of Materials

Schematic	- a basic map of what we are building
Arduino	- brains
Leds	- lighting
Resistor	- prevent burnout
Solder & Wire	- the glue that binds everything
PCB board	- to mount cube
Connectors	- to connect shield to Arduino

Schematic

The schematic is a graphical representation of an electrical circuit. This diagram shows the relationships between each of the components with standardized symbols representing each of the components.

The schematic, unlike the actual circuit board it represents, has a bit more flexibility. It doesn't show the actual layout and all of the connections that is the output of the design process but just the relationships. This sometimes means that different pieces of the schematic are not drawn as physically connected but instead show the names of the different traces. Thus the engineer who looks at the schematic understands how the signal propagates in the circuit. This makes it quite easy to read but even so it should represent all the pieces of the circuit.

This circuit diagram, Illustration 96, should seem to be a bit lacking as there is no power or ground. The answer is that each of the column labels will be connected to pin on the Arduino. These pins will control the flow of power to the anodes of the LEDs that are connected to it. The output of each of these LEDs is connected to their respective layer.

Each of the layer labels will also be connected to an Arduino pin and will be acting as ground. When the column is set to output and turned on while the layer is set to input and then the flow of electricity will go through the LED and that one will light up.

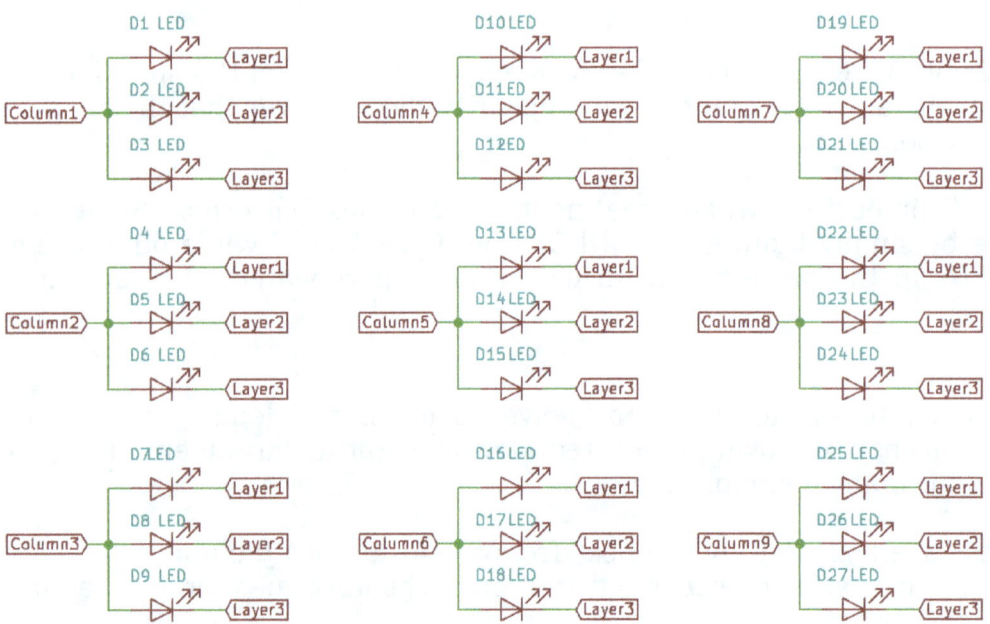

Illustration 96: 3x3x3 cube schematic controlled by Arduino

Assembly of cube

The Arduino powered cube is actually composed of two different printed circuit board. The lower layer is simply a standard Arduino. In this case the Atmega 256 compatible Arduino is quite overpowered for the task of powering this cube.

It was chosen because it wasn't very expensive but it had lots of pins and most importantly it has a lot of memory. Lots of memory will free us of the limitations of how many patterns we can create.

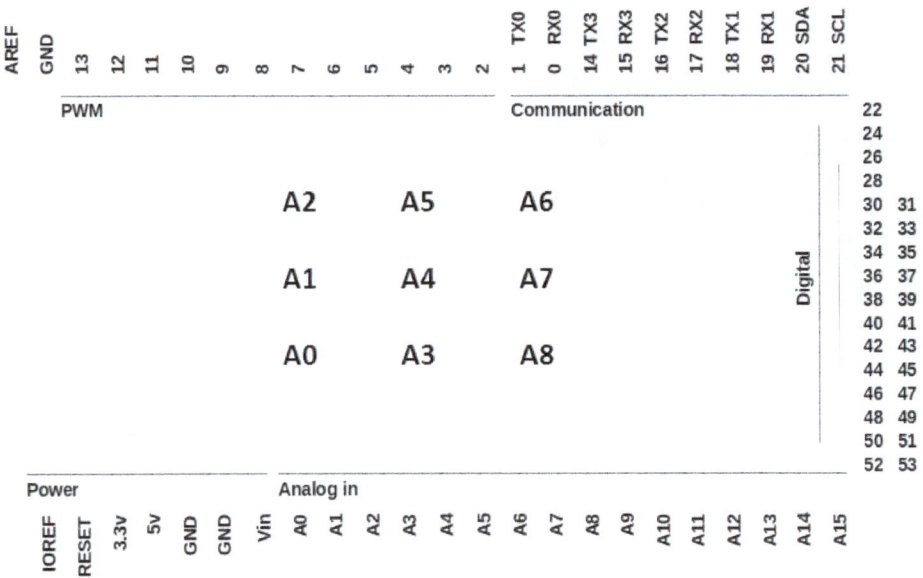

Illustration 97: Led cube showing pin mapping

The illustration 97 shows which pins are being used on the Arduino are connected to the various columns of the LED cube.

Illustration 98: ELEGOO ATmega2560

The Arduino pictured in illustration 98[1], shows all of the possible pins but it would be quite difficult to attach such a three dimensional cube to this device. What would be easier would be to attach the cube to a circuit board.

Such a circuit board is already available[28]. This printed circuit board is not only the same shape as the Arduino Atmega 2560 but it also pulls out each of the pins making it possible to create our own shield with access to any of the existing pins.

Illustration 99: PCB for Arduino MEGA 2560

This printed circuit board is rather small but it is just big enough that we can mount a 3x3x3 cube onto it and then connect the cube to the necessary pins so it can be controlled by the Arduino.

3x3x3 Cube portion
There is no right or wrong way to create the actual cube. While watching YouTube for researching I discovered quite a few different methods. All of these different methods incorporates some form of template to help keep the LEDs fixed while soldering them together.

Horizontal layers
Quite a few different methods seemed to create one complete layer at a time.

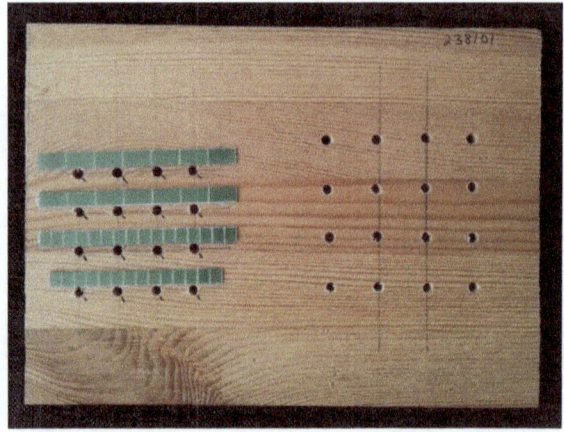

Illustration 100: A form for creating 3x3x3 or 4x4x4 layer

The most of the makers that I saw created this layer by bending the legs of the LEDs at right angles and soldering them together to both complete the circuit and form the cube structure.

I tried a variation of this technique by soldering straightened wire to the cathodes and bending the anode at a 90 degree angle. When assembling the layers, the anode was then soldered directly to the columns.

I did see yet another method for assembling a LED cube from Pollin which required the legs of the LEDs be bent into different patterns which when assembled would cause the anode to bend around the LEDs for the lower layers. This last kit was interesting to look at but the assembly seemed to be more difficult than the previous two other assembly methods.

Of these methods, perhaps using the LED legs as the structure may have been the simplest. It does pretty much guarantee the distance between LEDs and doesn't require that you create an absolutely straight wire for soldering LEDs together.

Note: Test as often as possible. If a cube is built using one of these methods it can be difficult to reach into the center of the cube to resolder any joints.

Vertical planes
Soldering together horizontal planes is actually very easy and does create a very stable structure. The only downside is it can be tricky to lower a complete plan down so it correctly aligns with the column when soldering it together.

This particular problem goes away when soldering together vertical planes.

The concept is quite similar to creating horizontal layers but then each plane is simply slotted into place. Once all planes are soldered into position you simply need to connect each level up by soldering a wire across each of the layers. This might not be quite as sturdy as the other methods but a LED cube is more of a visual item and not so likely to be touched.

Wiring up the cube
The Arduino provides all of the electronics so the only actual task is to connect up each of the Arduino pins to the columns and layers of the cube. The actual mapping of the cube columns with the Arduino pins are describe in illustration 97.

The only thing left to do is to solder a small wire between the cube leg and the appropriate Arduino pin, Illustration 101.

Illustration 101: Cube pins soldered to the Arduino pins

The red wires connect the 3x3x3 columns to the Arduino pins and the uninsulated (silver) wire with a resistor connect each layer to its appropriate pin. All of the wiring was one on the underside of the printed circuit board to help make the cube stand out without any other connections causing a visual distraction.

This method of wiring did work well and does work fine but only once it was all done did I realize that the resistors or the uninsulated wire could short out against the programming port on the Arduino. I simply provided an insulating layer between the shield and the Arduino when attaching the two.

Code the patterns

Computers are actually notorious for being able to run programs no matter how obscurely they are written or poorly formatted they are. Creating well named variables and data structures makes it easier for people write programs that work as well as reducing the number of bugs that are introduced.

To make the task of creating patterns as easy as possible I have "logically" laid out the columns using the diagram from illustration 102.

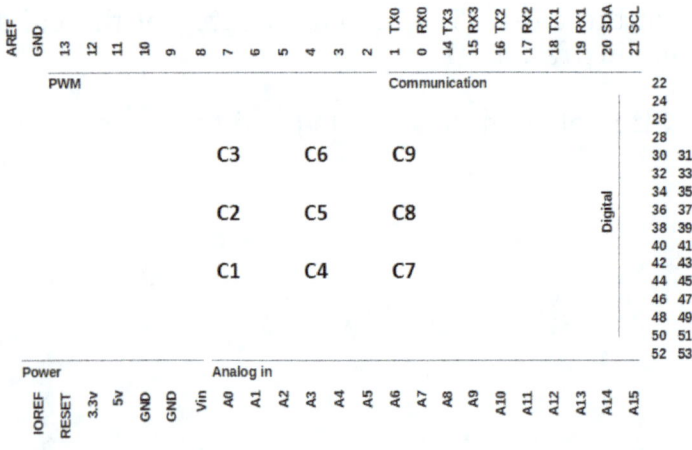

Illustration 102: Showing the names of each column

Good code should be like reading a book. Code should be in small blocks with each one dedicated to a specific task. I created new and more meaningful names for each column and I have slightly moved the pins around so the columns are consistently laid out.

```
int column1 = 54;    // analog pin 0
int column2 = 55;    // analog pin 1
int column3 = 56;    // analog pin 2
int column4 = 57;    // analog pin 3
int column5 = 58;    // analog pin 4
int column6 = 59;    // analog pin 5
int column7 = 60;    // analog pin 6
int column8 = 61;    // analog pin 7
int column9 = 62;    // analog pin 8

int bottomLayer = 41; // digital 41
int middleLayer = 43; // digital 43
int topLayer    = 45; // digital 45
```

Just like a graphics application the most important step is the ability to create a single dot on the screen. Once you can create a single dot you can then create the necessary routines for creating lines and other graphical elements.

This same logic also applies for creating patterns in a LED cube. The Arduino libraries already have methods which can be used for setting a pin for either input or output. Clever use of these methods allow us to provide power to a LED from one of the pins and cause a second one perform as ground.

The method that is used for selecting a pin for input or output is the "pinMode" method. It is necessary for setting a pin to OUTPUT in order to provide power. This is actually fairly intuitive and this same method is used for using a pin as ground. In order to set a pin to be ground you also set the pin to be OUTPUT, this is somewhat less intuitive.

This is perhaps best seen in a small example.

Example

```
void setup() {

        pinMode(54,OUTPUT); // provide power
        pinMode(41,OUTPUT); // act as ground

}

void loop() {

        // turn on a connected led
        digitalWrite(54,HIGH);              // connected to anode/positive
        digitalWrite(41,LOW) // connected to cathode/ground
        delay(2000);

        // turn off a connected led
        digitalWrite(41,HIGH)
        delay(2000);
}
```

This example would cause a LED to cycle on and off every two seconds. Either of these pins could be changed from HIGH to LOW to prevent the current from flowing.

Not all patterns are complex and can be created without much effort at all.

```
void floatingLayers()
{
  // set up all columns to use the current power level
  int idx;
  for (idx = 1; idx <= 9; idx++)
    digitalWrite(columns[idx],POWERON);

  for (int iter = 0; iter < 5; iter++)
  {
    Serial.println(iter);

    // turn off all, except for bottom layer of cube
    layersOff();
    digitalWrite(bottomLayer,LOW);
    delay(1000);

    // turn off all, except for middle layer of cube
    layersOff();
    digitalWrite(middleLayer,LOW);
    delay(1000);

    // turn off all, except for top layer of cube
    layersOff();
    digitalWrite(topLayer,LOW);
    delay(1000);

    // turn off all, except for middle layer of cube
    layersOff();
    digitalWrite(middleLayer,LOW);
    delay(1000);
  }
}
```

This pattern basically turns on all columns and it enables each level one at a time causing the LEDs from that level to light up.

This particular pattern is trivial and does not require using persistence of vision or any fancy tricks.

One of the patterns I implemented was a rotating layer. All the LEDs in the shape of a plane rotate around the X axis. This pattern was not terribly difficult but because it requires lit leads on multiple layers. Because of the construction of the cube with each layer acting as a common ground it is not possible to have lit LEDs on two levels that are not one above each other. Thus it is necessary to turn on each LED in the pattern briefly to display the entire pattern. In order for the pattern to be visible it may be necessary to lite up the LEDs multiple times.

```
void XAxisRotate_2() // uphill
{
 int pause = 2;
 for (int idx = 0; idx < 5; idx++)
 {
   columnOn(column3);  groundLayer(topLayer);
   delay(pause);
   columnOff(column3); unGroundLayer(topLayer);

   columnOn(column6);  groundLayer(topLayer);
   delay(pause);
   columnOff(column6); unGroundLayer(topLayer);

   columnOn(column9);  groundLayer(topLayer);
   delay(pause);
   columnOff(column9); unGroundLayer(topLayer);

   columnOn(column2);  groundLayer(middleLayer);
```

```
    delay(pause);
    columnOff(column2); unGroundLayer(middleLayer);

    columnOn(column5); groundLayer(middleLayer);
    delay(pause);
    columnOff(column5); unGroundLayer(middleLayer);

    columnOn(column8); groundLayer(middleLayer);
    delay(pause);
    columnOff(column8); unGroundLayer(middleLayer);

    columnOn(column1); groundLayer(bottomLayer);
    delay(pause);
    columnOff(column1); unGroundLayer(bottomLayer);

    columnOn(column4); groundLayer(bottomLayer);
    delay(pause);
    columnOff(column4); unGroundLayer(bottomLayer);

    columnOn(column7); groundLayer(bottomLayer);
    delay(pause);
    columnOff(column7); unGroundLayer(bottomLayer);

  }
}
```

The method, XaxisRotate_2, is actually only a small portion of the pattern. The pattern is the plane rotating around the X axis and so this actually requires different planes. I have created a different method for each plane and so all it is necessary is to call each of these methods one after another to generate the illusion of a rotating plane.

```
void XAxisRotate()
{
 for (int idx = 0; idx < 5; idx++)
 {
  XAxisRotate_1(); // horizontal
  XAxisRotate_2(); // uphill
  XAxisRotate_3(); // vertical
  XAxisRotate_4(); // downhill

  XAxisRotate_1(); // horizontal
  XAxisRotate_2(); // uphill
  XAxisRotate_3(); // vertical
  XAxisRotate_4(); // downhill

  XAxisRotate_1(); // horizontal
  XAxisRotate_2(); // uphill
  XAxisRotate_3(); // vertical
  XAxisRotate_4(); // downhill

  XAxisRotate_1(); // horizontal
  XAxisRotate_2(); // uphill
  XAxisRotate_3(); // vertical
  XAxisRotate_4(); // downhill

  XAxisRotate_1(); // horizontal
  XAxisRotate_2(); // uphill
  XAxisRotate_3(); // vertical
  XAxisRotate_4(); // downhill

  XAxisRotate_1(); // horizontal
  XAxisRotate_2(); // uphill
  XAxisRotate_3(); // vertical
  XAxisRotate_4(); // downhill
 }
}
```

In comparison to some of the simple patterns which can be done with 30 lines of code a rotating plane takes a couple of hundred lines.

Arduino IDE
The Arduino IDE can compile and upload the program to the atmega2560 but it is important that you select the proper device in order for this to work.

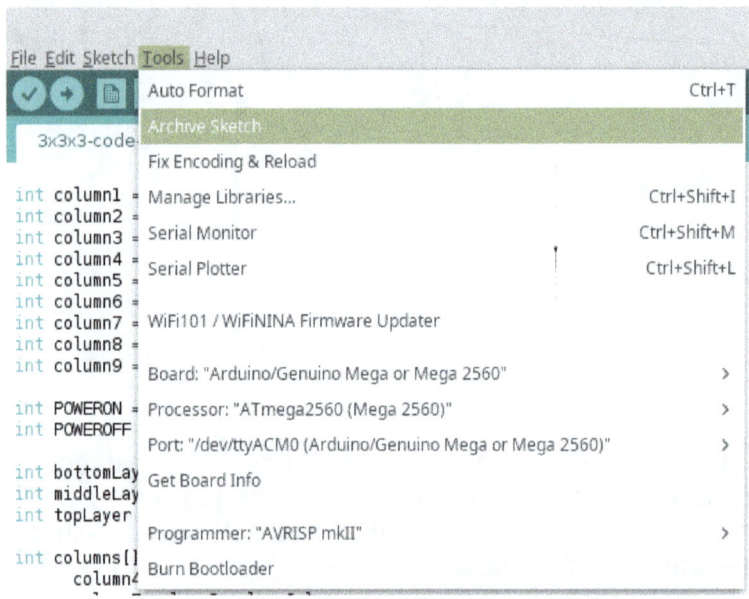

The Arduino IDE, version 1.8.9, actually selects most of the configuration, the only actual option that I had to select was the board. By selecting "Arduino/Genuino Mega or Mega 2560" caused all the other options to be defaulted to including the port. However, when using the Serial Monitor I did need to actively select the port which was correct but not selected.

Lessons Learned
Designing a LED cube or any project using an Arduino is fun and fairly easy. The Arduino has libraries that are designed to take out all of the complexities of programming such a device. There are methods for directly controlling pins, communicating via devices serially or other communication protocols such as I2C. This makes it possible to connect to other devices which support specific tasks which might be anything from a battery backed clock to a GPS device.

Despite how much functionality is unlocked and how easy they are to use it is highly advised that you experiment with them before launching into a project. Perhaps the best way is to use a breadboard and to test out individual parts controlled by the Arduino. This will help spot bugs or logic errors before any permanent changes (e.g soldering) are performed. It is possible to de-solder parts it will take much longer than to solder it in the first place.

I decided that I should connect all the LEDs together with wire that I had straightened. My wire was uninsulated and came in spools. I chose silver plated 0.8mm diameter wire, AWG 20 gauge. This is a pretty heavy duty wire which is makes a very stable LED cube. This wire is a bit thicker and so is more visible when examining the cube itself. I have also tried using 0.6mm diameter wire, AWG 22 gauge, which looks great but is a bit more fiddly to use as it bends quite easily when trying to assemble the cube.

The only problem with using wire is that it usually comes in spools which is the opposite of straight. There were a number of ways to straighten it but most of them are a variant of connect the wire to something and pull it straight. This is easily done with a pair of pliers and a vice grip.

The variety of LEDs that are available on the market is unbelievable. They litterally come in all shapes and sizes from 3mm to 1cm in diameter. The LEDs come in various styles of rounded (straw hat, bullet/rounded, flat top, inverted cone) as well as square. It wasn't the style of the LED but rather diffused that made the difference. Diffused LEDs will disperse the light in all directions which is especially important viewing the LED cube from different angles.

When creating a template make sure that the LEDs are not too close together. It is not usually too difficult to solder them in the template but the closer they are together the more difficulties there can be if you need fix a solder joint after the cube has been assembled.

Perhaps the most important lesson learned is to take your time. This is important when creating a template or when soldering. The solutions suffer when rushed or if you are not feeling calm and patient.

Elegoo Atmega 2560
https://www.elegoo.com/product/elegoo-mega-2560-r3-board-blue-atmega2560-atmega16u2-usb-cable/

Shield for Arduino atmega 2560
https://www.amazon.de/gp/product/B01FTWJR9I/ref=ppx_yo_dt_b_asin_title_o02_s00?ie=UTF8&psc=1

11 LED CUBE WITH CUSTOM PCB

In the previous chapter I created a 3D cube that was controlled by an Arduino. It didn't have to be an Arduino, it could have just as easily been a Raspberry Pi or any other small form factor development board.

When using a small form factor board this pretty much solves the problem of all the necessary electronics for controlling the cube. Usually these boards provide a convenient way to attach other electronics. In the case of the Arduino it was possible to create our own shield with the actual cube electronics mounted on it which was ultra convenient.

The Arduino is a neat platform as it provides a proven micro-controller board, an integrated development environment and support libraries for many common functions. There is almost nothing not to like about it except the cost. These devices are affordable for development or prototype projects but could be considered fairly costly for a small single purpose device such as a custom build networked set of thermometer type devices for each room of your home.

A lot of the parts that make up a controller board are themselves actually not that expensive. Through hole versions of these parts could be purchased and soldered to a prototyping board or you could even design your own custom board.

Creating your own printed circuit board could in theory solve two different problems. The actual components for custom solutions may be a lot smaller, cheaper but also easier to wire up.

 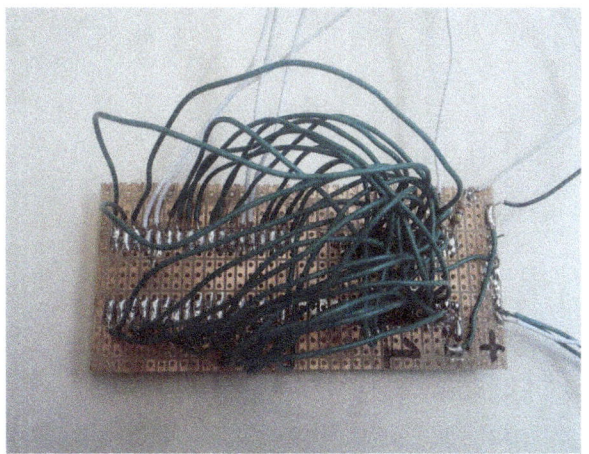

Illustration 103: First self designed - front *Illustration 104: First self designed - back*

Printed circuit boards encapsulate all of these same wires but in a way that is both visually appealing and is more durable. This seven segment display was a small test project but never went anywhere due to the problems of connecting all of the wires and keeping them from becoming disconnected when moving the device.

The good news is that creating a custom PCB has never been easier than it is today. In the past some of the barriers preventing you was dealing with foreign currency, transport of the layout to the manufacturer, transport of the boards to you, as well as potentially facing foreign language issues.

Most of these are no longer a problem for quite a few different reasons. The electronics industry is so much more advanced today than it was back in the olden days. The rise of credit cards, wire transfers and paypal make it just as easy to transfer money between banks as between different countries. The methods of transport have not significantly changed in the last 40 years but the amount of goods have risen approximately 300% from 1992 to 2012. The improved delivery infrastructure whether it is email, ships or trucks will help not only large companies but also the home hobbyist.

In this chapter I am going to actually create my own board and solder both the parts and the cube to it. Finally, I will be programming it with a few patterns.

The Arduino that was used to create a cube was powered by an Atmel ATmega2560, but I will use a much less powerful Atmel 2313 for my custom board. It has enough pins to fully control the 3x3x3 cube without the need of adding an I/O expander or using a clever design such as charlieplexing. This allows us to focus more on the board we are creating and less effort for complicated electronic issues.

Just like in chapter 10, the individual LEDs in the cube will be powered via the column and each layer will be connected together and then used as the common ground. The previous solution was feeding the power to the different LEDs directly from the Atmel and back to ground which essentially made the micro-controller the "powerplant" for the cube.

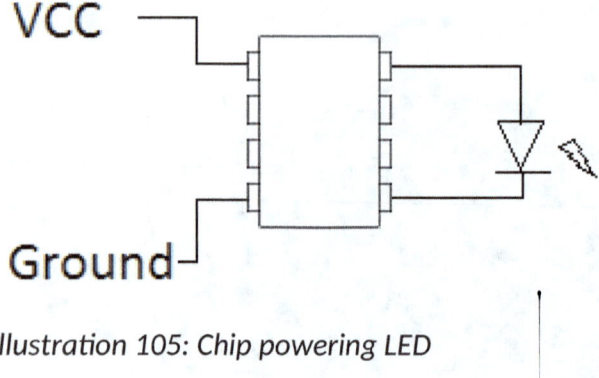

Illustration 105: Chip powering LED

This limits the amount of LEDs that can be on at any one time due to the power limitations of the chip. This is true not only for Atmel but almost any chip on the market. It would be difficult, for example, for a chip to directly power a large motor.

One possible pattern for the cube would be to switch on all LEDs for a level but another would be to switch on LEDs for all levels at the same time. Considering the limitations on how much power can be provided by the chip this would put stress on the micro controller as well as affect the brightness of the LEDs. The brightness of the LEDs will vary based on the amount of power available which would vary based on the number of LEDs switched on at any point in time.

The goal of my own PCB cube is to move the micro-controller into a controlling position where it is not sourcing the power but is only controlling the power flow. In order for that to happen we need to add a "power switch" that can be controlled by the micro-controller. A common switch for that task would be the NPN transistor.

The NPN was described in chapter 1, but essentially it is a switch that will connect the collector with the emitter when the base receives power. If the base receives greater than 0.7v then it will "turn on" and release the flow from the collector, see Illustration 106.

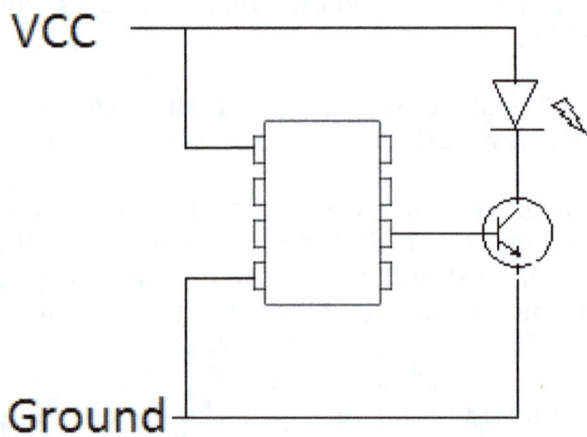

Illustration 106: Chip controlling power to LED

The only real change to the logic of our cube is that the micro-controller needs to control the NPN not the LED leg. This might sound tricky as now for each LED column there will be one additional trace going to each LED – actually only for each column.

Using the NPN as a switch for powering the columns is simple. The power is connected to the collector while the micro-controller is connected to the base. When the power on the base exceed the threshold then the power flows from the collector to the emitter and then on to power the column of LEDs.

Just like the Arduino solution, it is not enough to control only the power but also the ground. If all layers go directly to ground then each time a LED column is powered up that LED will light up on all three layers. To prevent this the micro-controller will also need to control each layer.

Controlling the layers is done in almost the exact same way as for controlling the columns. The difference is that this NPN is being used to connect through to the ground. Each layer is connected to the collector and the micro-controller is connected to the base. When the power on the base exceed the threshold then the power flows from the collector to the emitter and then on to ground which completes the circuit.

The schematic
The schematic for my printed circuit board cube.

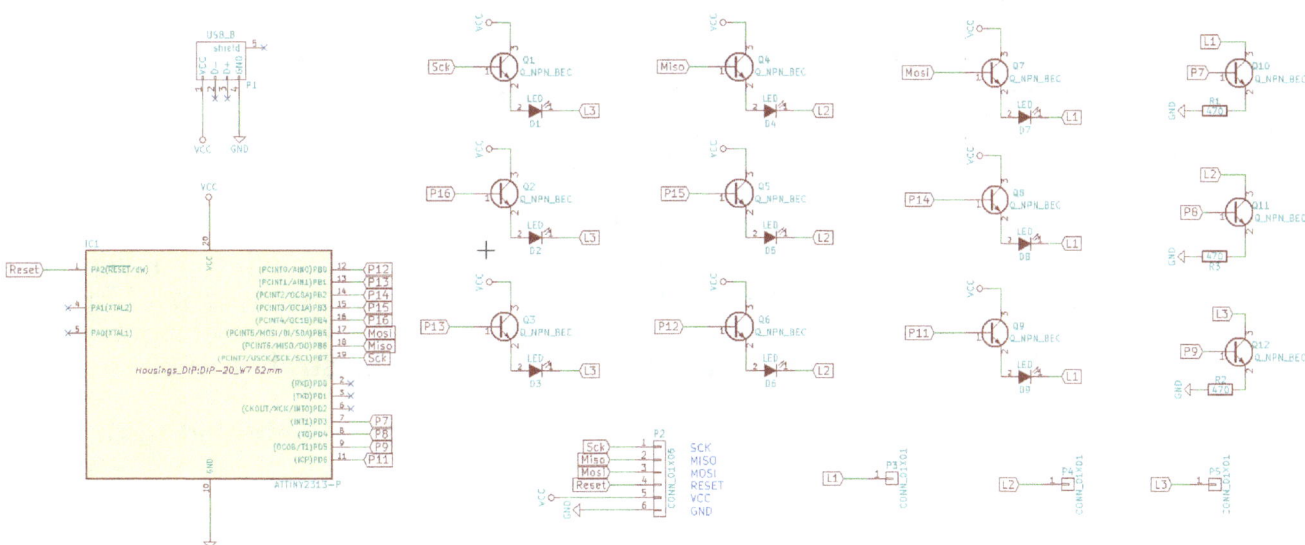

Illustration 107: 3x3x3 cube schematic

The reason this schematic is so spread out is that it is heavily using net names. Net names can be used in order to make a schematic more legible. Without them there would be many lines crossing all over the diagram which would make reduce the readability and thus the value of the diagram.

These net names are virtual connections on the schematic that will be physically connected on the circuit board. A net consists of two or more labels that creates the connection even though no actual line between the parts is displayed.

> LB16

    net name LB16

There no limit to the number of connections to a net. The minimum is two connections which may be used to connect two different pins but a net can support many different connections. One example of this might be the VCC net which is used in a number of different locations to provide power to the columns.

In my example cube, not all of the net names are as meaningful as they could be. Some of the more meaningful tags in this schematic are used for the header pins to the chip.

- MISO
- MOSI
- RESET
- SCK
- VCC
- GND

Some of the tags unfortunately are using the much less meaningful net name of P15 instead of a more meaningful name such as "Column 5".

I should have renamed the nets to more friendly names but as this doesn't actually add any new functionality and because this is a fairly time consuming process it didn't get done. Just like in software development, selecting good names at the beginning of a project makes things easier and then will not need to be "fixed up" at a later date.

It might not be immediately obvious from looking at the schematic but you may have noticed it is actually a bit unusual. The schematic actually does not seem to be displaying all the parts for this circuit. We can verify that this is the case because a 3x3x3 LED cube must have 27 LEDs but the schematic only displays 9 of them. Each of the columns on the schematic, Illustration 107, should be changed to look similar to illustration 108.

This way you can see the power going to each anode and each cathode tied to the layer which is in turn then attached to ground.

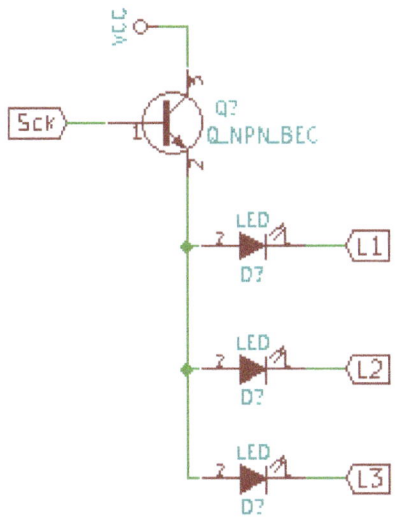

Illustration 108: Proper schematic for the column

The reason the rest of the LEDs are not listed on the schematic is that they do not actually provide any further information for the board manufacturer. Not only don't they provide any further information they actually will not be mounted on the PCB.

A LED cube is a fairly unique type of electronic circuit as none of the LEDs are actually mounted on the board itself. The LEDs are connected together in a three dimensional lattice connected to the circuit board. The frame connects the LEDs in an interesting way and then attach a few of their legs to the board so they can be individually controlled.

PCB Layout
It is an art form to convert a schematic into a circuit board as there are many correct yet different methods for any layout and not all designs are equally good. Multiple designs can do the same thing, but some may have heating problems, electrical interference issues or may even require too large of a PCB. The reason that the design of the circuit board is tricky is that the various traces cannot cross each other.

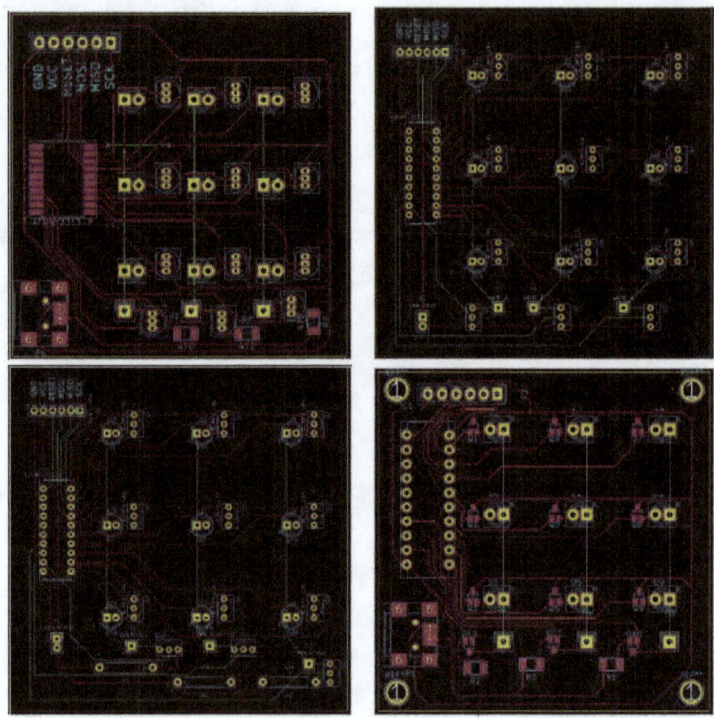
Illustration 109: 3x3x3 attempts at control boards

There are several different ways to get around this restriction. The simplest example of this is to have traces on both the front and on the back sides of the circuit board. Because each of these layers are insulated from each other, the traces do not physically cross each other.

The printed circuit board in Illustration 110 only uses one side. More complicated circuit boards such as computer motherboards can consist of six or more layers and some CAD tools are capable of designing up to 32 layer circuit boards.

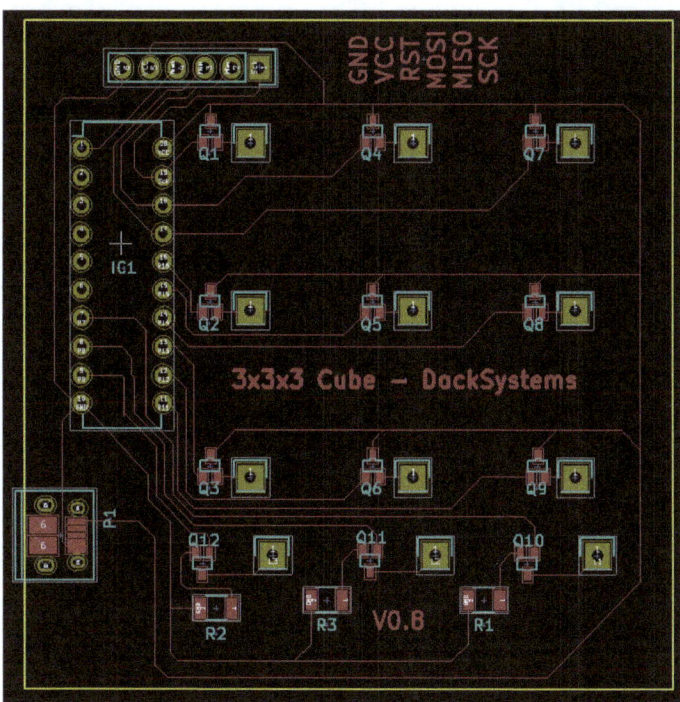
Illustration 110: 3x3x3 printed circuit board

A second method would be to either doing a lot of planning up front or to change your design while doing the layout. Micro-controller's do have quite a few different pins and in many cases these pins offer the same functionality. It is as you get closer to the final design of the circuit board you can see how slight changes to your schematic may simplify the layout of the board.

For smaller projects we can simply modify the schematic and resume with the design of the PCB layout. Doing this makes the development of the printed circuit board an iterative process.

Fabricating the circuit board
Telling a story to a friend is easy but to tell it in such an unambiguous way and to do it so you both can agree on all of the details is difficult. This is a problem that you do not want to have when you decide to fabricate your board. You want the manufactured result to be identical to what you are expecting.

In order to ensure there are no ambiguities it was decided upon a long time ago for a standard format for the transmission of this type information. This format was the Gerber file format.

What are Gerber files?
Gerber files are a defacto industry standard format for describing printed circuit boards. Gerber files were originally designed by the Gerber Systems Corporation to drive their vector photo plotters back in the 60's and 70's. This process was quite similar to what was needed for the creation of printed circuit boards and has been used ever since.

This "format" is an ASCII text file for representing vectors of two dimensional binary images. When people refer to Gerber files for the creation of a PCB it usually does not refer to a single file but several files of this format that when taken together represent the complete picture of the board from copper layers to silk screens. This collection of Gerber files usually also contains a drill file which then provides the final information required to create a three-dimensional representation of the printed circuit board. The collection of Gerber files are usually zipped together and sent to the board fabricator in a single archive.

Like any format, the Gerber format has changed over time. The current version is from 2014 and is sometimes referred to as version RS-274X. This latest format includes additional meta data. When this additional data is available then these files are called X2 while files without this data are X1 files. The previous format is version RS-274D.

Your mileage may vary
A lot of the difficult work has already been achieved when you are ready to get your board fabricated.

>   Have an idea
>   Create a schematic
>   Build a prototype
>   Select source for your components
>   Create a printed circuit board layout

Any of these steps can keep you busy for days until you have completely solved all related problems associated with a particular step. The final step of preparing for fabrication should be and is virtually trivial. Simply generate the Gerber files. Yet, there is still a step that should be inserted before creating your Gerber files. The step is to compare your PCB design with the limitations of your manufacturer.

Ideally these limitations should be kept in mind before designing your board as they may have an impact on your layout. Here are just a few of the limitations that need to be considered when picking a manufacturer.

>   Maximum number of layers
>   Board material
>   Max board dimensions
>   Board thickness
>   Min trace width
>   Min trace spacing
>   Min via hole size
>   Min and Max drill hole size
>   Min character width
>   Min character height

These considerations are not only what do you need for your board (i.e. 17 layers) but can your manufacturer create a board with such specifications. Some of these, such as the size of characters on your silk screen is a cosmetic issue while others such as via hole size or trace spacing may prevent you from fabricating your circuit board. Thus, it is imperative that you check the capabilities of your manufacturer to make sure that nothing on your layout conflicts with these capabilities.

More information on how to do various tasks in Kicad as well as techniques are described more in the appendix.

Picking a Manufacturer

This is actually a tricky task. Just like purchasing anything else a certain amount of investigation should be performed. This can be a simple as a quick internet search for "PCB manufacturers", "PCB manufactures cheap", or perhaps even "PCB manufactures USA".

There are really not a lot of questions that unless you have a reference to speak with.

- How long have they been in businesses
- What certifications do they have
- Where are the boards manufactured
- Turn around time
- Cost
- Minimum order
- Are they a manufacture or just a broker

All of these are important questions but perhaps the most important is the last one in the list. When I first started looking for a board manufacture I didn't know that there existed brokers who would be happy to play the middle man which is their only role.

Of course you might also do some internet searches to see if there are any happy or unhappy customers before deciding to order from a specific manufacturer.

When I picked my manufacture I chose one that was fairly cheap but also one that seemed to understand that some of their customers may be hobbyists who are less familiar with the entire process. How they did this was by having notes on their website explaining how to export your information from your CAD tool. This wasn't a single tool but actually a list of instructions for quite a few common CAD tools on the market.

Other manufactures provided other services such as accepting the CAD project instead of gerber files and they would extract the information necessary for creating the PCB's. I even saw one site that provided their own CAD software. You use their software which is already closely aligned with the manufacture which streamlines the ordering process.

Assembly

The assembly is essentially just the soldering of the parts together.

1 USB mini port
Atmel 2313
20 pin socket

27 LEDs
12 transistors
3 resistors
print circuit board

The easiest part of the soldering process was soldering the parts onto the circuit board. Simply start in the middle of the board with the smallest parts first and work your way out.

Illustration 111: 3x3x3 board with all parts soldered to it

Despite the size of some of the surface mount parts this is the easiest part of the assembly process. Each part has either a pad or through hole so it can be attached to the board. The real tricky part is building the three dimensional cube built out of LEDs.

In the previous chapter a LED cube was built by creating panels of LEDs. When building this cube I have created layers instead of panels.

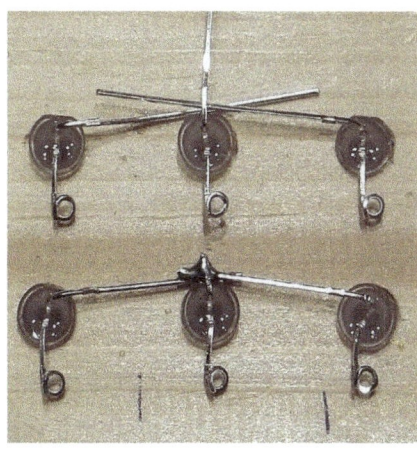 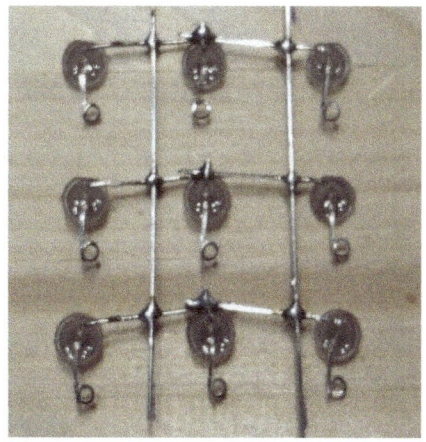

Illustration 112: Partial layer   Illustration 113: Full layer

Each anode was bent into a small loop while the rest of the cathodes were soldered together. In addition to those cathodes some additional wires were used ensure that all cathodes were connected to each other. This will allow the entire layer to have a essentially a common cathode.

The process of assembling these layers is just a matter of carefully threading the wires from each column through the loop which is on the end of each anode.

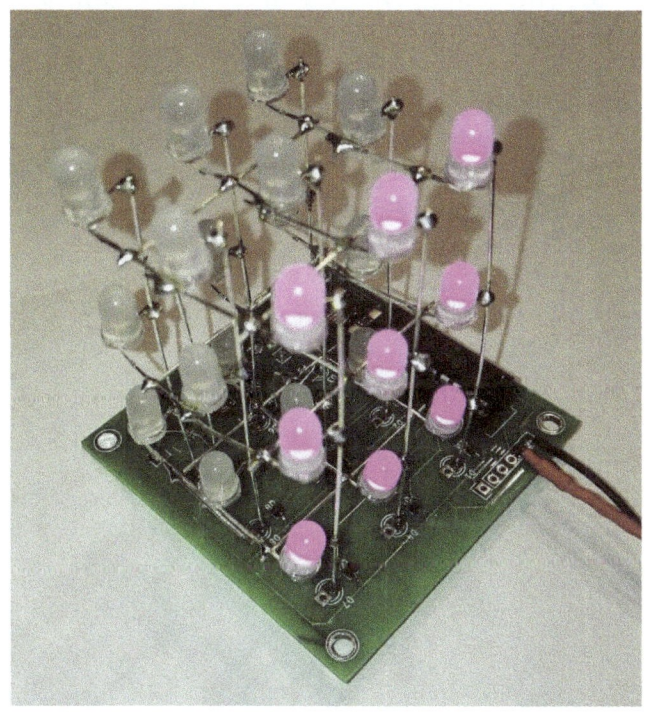

It might not be immediately obvious but this cube does not actually use a mini USB connector. Due to difficulties this connector was abandoned. You can see in this picture that the power and ground for the programming pins can also be used to power the cube.

Software
There is a very big difference in the software that drives the cube.

It simply cannot be argued that the Arduino does not have a rich and easy to use software development environment.

Arduino cube initialization
```
void setup()
{
  int idx;
  // setup all columns for output
  for (idx = 1; idx <= 9; idx++)
    pinMode(columns[idx],OUTPUT);

  // setup all layers for output (e.g as power sink)
  pinMode(bottomLayer,OUTPUT);
  pinMode(middleLayer,OUTPUT);
  pinMode(topLayer,OUTPUT);
}
```

The pinMode method is used for properly setting a port for input or for output. I didn't bother to create either a procedure or a macro which would give me that same level of abstraction.

The port configuration in the Atmel chip 2313 is controlled by DDRB and DDRD. These are bit mapped values with each bit representing a different leg. The direction of the port is defined with a zero for the bit for input and a one for output. The LED cube will only control the LEDs and not receive any input so all of the macro values that I have defined are for setting an individual value.

My 3x3x3 cube initialization (partial)
```
#define COL1  PB7
#define COL2  PB6

unsigned char COL1mask = (1 << COL1);
unsigned char COL2mask = (1 << COL2);

#define COL9  PD6
unsigned char COL9mask = (1 << COL9);

#define LAYER1  PD3
#define LAYER2  PD4
#define LAYER3  PD5
unsigned char LAYER1mask = (1 << LAYER1);
unsigned char LAYER2mask = (1 << LAYER2);
unsigned char LAYER3mask = (1 << LAYER3);

void define_output_ports()
{
 // setup the columns as output
 // setup port B
 DDRB =(
        COL1mask |
        COL2mask |
        COL3mask |
        COL4mask |
        COL5mask |
        COL6mask |
        COL7mask |
        COL8mask
    );
```

```c
        // setup port D
        DDRD =(
                COL9mask |
                LAYER1mask |
                LAYER2mask |
                LAYER3mask
                );
}
```

In this initialization procedure I am creating mask in a very verbose way. It isn't necessary to define variables or macros in this way but it does make reading the code much easier to understand. More information about bit manipulation can be found in the Appendix C programming refresher.

Source code

```c
#include <avr/io.h>
#include <util/delay.h>         // for _delay_ms()

#define COL1  PB7
#define COL2  PB6
#define COL3  PB5
#define COL4  PB4
#define COL5  PB3
#define COL6  PB2
#define COL7  PB1
#define COL8  PB0

unsigned char COL1mask = (1 << COL1);
unsigned char COL2mask = (1 << COL2);
unsigned char COL3mask = (1 << COL3);
unsigned char COL4mask = (1 << COL4);
unsigned char COL5mask = (1 << COL5);
unsigned char COL6mask = (1 << COL6);
unsigned char COL7mask = (1 << COL7);
unsigned char COL8mask = (1 << COL8);

#define COL9  PD6
unsigned char COL9mask = (1 << COL9);

#define LAYER1  PD3
#define LAYER2  PD4
#define LAYER3  PD5
unsigned char LAYER1mask = (1 << LAYER1);
unsigned char LAYER2mask = (1 << LAYER2);
unsigned char LAYER3mask = (1 << LAYER3);

void define_output_ports()
{
 // setup the columns as output
 // setup port B
 DDRB =(
        COL1mask |
        COL2mask |
        COL3mask |
        COL4mask |
        COL5mask |
        COL6mask |
        COL7mask |
        COL8mask
     );

 // setup port D
 DDRD =(
        COL9mask |
        LAYER1mask |
```

```c
            LAYER2mask |
            LAYER3mask
            );
}

void sidePanels()
{
  int leftpanel = COL1mask | COL2mask | COL3mask;
  int midpanel = COL4mask | COL5mask | COL6mask;
  int rightpanel = COL7mask | COL8mask ;
  int idx;

  for (idx = 0; idx < 4; idx++)
  {
   PORTB = leftpanel;
   PORTD = LAYER3mask | LAYER2mask | LAYER1mask;
   _delay_ms(250);

   PORTB = midpanel;
   PORTD = LAYER3mask | LAYER2mask | LAYER1mask;
   _delay_ms(250);

   PORTB = rightpanel;
   PORTD = LAYER3mask | LAYER2mask | LAYER1mask | COL9mask;
   _delay_ms(250);

   PORTB = midpanel;
   PORTD = LAYER3mask | LAYER2mask | LAYER1mask;
   _delay_ms(250);
  }
  PORTB = 0;
  PORTD = 0;
}

void crazyPanels()
{
  int idx;
  for (idx = 0; idx < 4; idx++)
   {
    PORTB = COL1mask | COL2mask | COL3mask;
    PORTD = LAYER3mask | LAYER2mask | LAYER1mask;
    _delay_ms(300);

    PORTB = COL1mask | COL5mask ;
    PORTD = LAYER3mask | LAYER2mask | LAYER1mask | COL9mask;
    _delay_ms(300);

    PORTB = COL3mask | COL6mask ;
    PORTD = LAYER3mask | LAYER2mask | LAYER1mask | COL9mask;
    _delay_ms(300);

    PORTB = COL3mask | COL5mask | COL7mask ;
    PORTD = LAYER3mask | LAYER2mask | LAYER1mask ;
    _delay_ms(300);

    PORTB = COL1mask | COL4mask | COL7mask ;
    PORTD = LAYER3mask | LAYER2mask | LAYER1mask ;
    _delay_ms(300);

    PORTB = COL1mask | COL5mask ;
    PORTD = LAYER3mask | LAYER2mask | LAYER1mask | COL9mask;
    _delay_ms(300);
   }
}

int main(void)
{
  define_output_ports();

  while (1)
```

```
    {
      crazyPanels();
      sidePanels();
      _delay_ms(500);
    }
  }
```

Lessons learned
The design of the cube actually went quite well. However, it was designed and built based simply on the parts that I could find in the CAD software. The design of and the creation of the board went flawlessly. Yet in retrospect I would have made a few different choices.

Power
I chose a USB mini port to power the device. This was because the part was reasonably large, provided a firm connection and would allow the ability to purchase any USB cable to power the cube. The idea was good but the solder pads for this part were almost entirely underneath the power jack. I did manage to successfully solder this part one in six attempts.

Later I did manage to find one part that was a USB mini port with through hole pins. That would have been a more appropriate choice even if I had to create a footprint for this device myself. A second choice would have been to select a power port in barrel form. This would have also allowed for a reasonably standard power cable but one that is easier to solder.

Debugging
When designing a board perhaps some thought should be given to also adding some testing points. Testing points would have helped when trying to decide if I had a problem with my USB power port.

NPN transistor
My experimental design to use a NPN transistor worked just fine. The only problem with this choice was the actual footprint was tiny. I never actually examined the part before the board was designed. I simply designed the board using the CAD software and this allowed me to zoom in when I needed to. It is possible that this NPN was actually a good size but next time I should have more attention to the footprint of the various components.

Software Upgrade
During the process of designing the board I happened to upgrade my laptops operating system. When setting up my laptop I decided to install the latest version of the CAD software. The software did indeed have improvements but it actually did slow me down due to problems with footprints. It was unfortunate that one of the major changes was in how footprints were handled.

In the future I would refrain from switching software versions in the middle of the project.

12 "C" PROGRAMMING REFRESHER

This chapter is just brief description showing the differences between python and C. It is intended as crutch to assist anyone with some development experience in another language to translate that knowledge into C. This chapter does not attempt to be a complete tutorial on C, some subjects such as pointers are only briefly examined, while others such as I/O are not covered at all.

A programming book on a single language could easily be hundreds[48] of pages long and that level of knowledge won't be necessary to get started with the Raspberry Pi or Arduino.

Interpreted vs Compile
These days computer programs are high level languages that have been created to increase programmer productivity. High level languages are almost English like which describe what is to be done. These languages can be either interpreted or compiled into machine code.

Interpreted languages such as python have a program called an interpreter running which reads through the program one line at a time and will execute the statements. This is less efficient as the program will be scanned and evaluated each time it is run. In some cases, such as loops, some lines may be evaluated multiple times per run.

Just like anything else, there are trade offs for interpreted programs. Time saved by not having a compilation step may speed up development however syntax errors may not be discovered until sometime during runtime. This is not a problem during testing but less desirable in production. Like in any other language, the discovery of syntax errors or bugs starts the cycle of fixing and testing until the program is perfect.

In a compiled language the compiler converts the program into object code which is a language that the computer can understand and execute directly. During this conversion syntax is checked, verifying variables and functions are defined and that the proper data types are being passed in the function calls. This isn't to say that the compilation process makes the programs less buggy, it doesn't, but it does guarantee that it is syntactically consistent and doesn't have any typo's in the variable or function names.

In order for either the compiler or the interpreter to understand the program, the language must be defined with a given syntax. This syntax uses special words or symbols to mark the end of the statements. Once the interpreter or compiler interprets the code, it can be

---

48 O'Reilly Media: Practical C++ Programming, 2nd Edition. 576 pages

further acted on, either by evaluating the next command or outputting the low level statements that the computer can directly understand. In C the symbols are semi colon, colon, left and right curly brace, but in python this is tabs and colons

Both interpreted and compiled languages essentially are comprised of the same types of elements such as variables, methods or functions, operations and flow control statements. Learning the syntax of a new language might be done in anywhere from a few hours to a day. Usually the problem when learning a new language is not the syntax but learning the functions, classes and libraries that assist in the software development process.

Variables

As python is an interpreted language the variable assignment is not statically typed (e.g checked at compile time). The interpreter knows the data type based on the value that has been assigned to it. There is nothing preventing variables from being reassigned values of a different type.

```
#!/usr/bin/python

a=10
b=10
c=a * b

print a
print b
print c

c="house"
print c
```

Software written in C is statically typed. This means that all variables need to be defined before they are used and that all variable assignments are done using the proper data type or compatible data types. Thus it is possible to assign a integer values to a floating point variable in C but it is not possible to assign a string to a numeric variable.

In C, there are a lot of data types and some additional key words which will alter what types of data they can hold. For the present, only the following types will be discussed; do not be lulled into believing this is a comprehensive list of data types for C.

Type	Data
char	Holds a number which can be stored in a 8bit variable. This may be from 0 - 255 which can be displayed as character, or a value in the range of -128 to 127
int	Positive and negative number
long	Positive and negative number
float	Positive and negative fractional numbers
double	Positive and negative fractional numbers
void	Nothing. This is not used when passing in parameters to functions but rather when functions will specifically not be returning a value.

It is hard to talk about the maximum value that any of these variables can hold as C allows variables to contain either only positive numbers (unsigned) or positive and negative numbers (signed) and when the compiler is not told explicitly what to use it will make its own determination.

This is usually not a problem and by default the compiler will probably create signed variables. It is usually better to not let the compiler do your decision making but to simply be careful when declaring your variables.

```
#include <stdio.h>

int main(int argc, char **argv)
{
    char building_age = 208;
    signed char depth_of_hole = -15;
    char depth_of_pothole = -15;

    printf("cathedral is %d years old\n",building_age);
    printf("hole in ground is %d units\n",depth_of_hole);
    printf("pothole is %d units\n",depth_of_pothole);
}

pi@raspberrypi ~/my_examples/c-ex $ gcc one.c -o one
pi@raspberrypi ~/my_examples/c-ex $ ./one
cathedral is 208 years old
hole in ground is -15 units
pothole is 241 units
```

Otherwise, one day the program will encounter a border case where the variable type matters and the results may not be what you expected.

Outputting formatted data in python is actually quite similar to C yet not exactly the same. In python it is not necessary to declare your variables before using them but it still is important to know what the type of data is when trying to output them.

Data type	Python Format	C Format
integer	%d	%d
long		%ld
floating	%f	%f
double		%lf
string	%s	%s

Just like C, the format strings contain a percentage sign with a letter to indicate the variable type but the syntax for passing the variables to the print statement is slightly different. In C the print statement is actually a function call which is just like any other function call.

```
printf ( "format string", value1,value2);
```

Printing in python is less like a external function call and more like a built in keyword[49]. To pass parameters to the print statement simply pass a format string, a percent symbol and one or more parameters.

```
print "format string" % singlevalue
print "format string" % (value1, ..., valuen)
```

There another major difference between the print and printf statements is that in python each print statement will automatically output a carriage return line feed (crlf) to advance to the next line. This must be done explicitly in C.

---

[49] Actually in python 2.x the print statement is a keyword but in 3.x print is a function. The version of python on my Raspberry Pi is 2.7.3

The two languages are quite similar as far as the syntax goes if you can overlook that one is interpreted and the other compiled. The first real big difference between the two is strings, or the lack of strings in C. This doesn't mean that C cannot have what normal people would call strings but as a developer a bit more work is involved.

In C, instead of a string there are arrays of characters with the last one in the array containing the value NULL (zero) which marks the end of the string. This is not meant to imply that only the last entry in the array will be a zero, a zero can exist anywhere in the array and that will signify the end of the string.

Type	Data
char mystr[10]	An array of 10 characters with an index from 0 to 9
char mychar	A single letter

There are functions which can be used for working with these arrays of characters and treating them like strings. It is possible to use these functions to manipulate these arrays in similar ways to standard strings in other languages. One of these functions, strcpy, will perform the assignment of string data to character arrays.

Type	Example	Data
char mystr[7]	strcpy(mystr,"chris")	mystr[0] = 'c'
		mystr[1] = 'h'
		mystr[2] = 'r'
		mystr[3] = 'i'
		mystr[4] = 's'
		mystr[5] = 0
		mystr[6] = "undefined"
char mystr[7]	strcpy(mystr,"chris");	mystr[0] = 'b'
	strcpy(mystr,"bob");	mystr[1] = 'o'
		mystr[2] = 'b'
		mystr[3] = 0
		mystr[4] = 's'
		mystr[5] = 0
		mystr[6] = "undefined"

It may seem somewhat elementary, but each array must be long enough to contain the "string" value plus one additional value at the end. This final entry will contain the value zero. The reason for this is that array's are simply a small block of memory with all the values adjacent to each other. There is no hidden array length that is available to help with displaying the contents of the array.

Here is some pseudo code for copying a string in C.

```
index = 0
while source[index] <> 0
begin
        destination[index] = source[index]
        index = index + 1
end
destination[index] = source[index]
```

Nobody would want to do this every time just to copy a string so of course there are a number of functions available to support C strings, but they depend on the strings having a zero at the end, which is referred to as being NULL[50] terminated. Three of the many functions for manipulating character arrays are strlen, strcpy and strcat.

strlen	returns the length of the array up to the first NULL
strcpy	copies a character array from one to another
strcat	concatenates a character array from one to the end of another

Granted using these functions may not be quite as easy as direct string assignment or string addition but in general they make it almost as easy to use character arrays as an actual string data type.

It is very important that strings are NULL terminated, otherwise there will be very bad results. The reason being that without the NULL termination these helper functions will carry on a random amount through memory until a NULL is found. No matter what function is being used, the resulting behavior will probably not be the desired outcome.

Below are a couple of programs to briefly show the differences between C and Python in string handling.

Python
```
#!/usr/bin/python

firstname="santa";
lastname="claus";
fullname=firstname + " " + lastname;

print fullname
```

C
```
#include <stdio.h>
#include <string.h>

int main(int argc, char **argv)
{
  char first[50];
  char last[50];
  char fullname[100];

  strcpy(first,"santa");
  strcpy(last,"claus");

  strcpy(fullname,first);
  strcat(fullname," ");
  strcat(fullname,last);
  printf(fullname);
  return 0;
}
```

Order of operations

When evaluating equations it is important to keep in mind the order of operations. That is, some operators will be evaluated first before others. For general math there are only a few operations and it is easy to have an overview of them.

Order	Operator	Description
1	( )	Parentheses
2	* /	Multiplication and division
3	+ -	Addition and subtraction

---

[50] A NULL is defined as 0. In some header files it may be defined as ((void *)0) but anywhere a NULL can be used the constant 0 can also be used.

In programming, certainly in C, there are more operations and some of them may have a somewhat counter intuitive ranking – such as bitwise shifting.

Order	Operator	Description
1	( )	Grouping
2	! ~ - + * & ++ -	Unary operations
3	* / %	Multiplication, Division, Module Division
4	+ -	Addition and subtraction
5	<<	Bitwise shift left (multiplication by a factor of two)
		Bitwise shift right (division by a factor of two)
	>>	
6	&	Bitwise and
7	^	Bitwise exclusive or
8	|	Bitwise

The safest bit of knowledge is that you can always use parenthesis to force a specific order of evaluation.

Standard math order of operations also apply for C programs and in our small sample a + b * c - d equals 41 for both C programming as well as traditional math homework.

```
a     = 3
b     = 5
c     = 8
d     = 2

val   = a + b * c - d
      = 41
```

Simply adding a few parenthesis allows us to slightly modify the outcome of the equation.

```
val   = (a + b) * (c - d)
      = 48
```

The order of operations is not different for the math you use day to day but when looking at the some of the bit operations in C the entire picture changes. It may seem like bitwise shifting either left or right should have a similar precedence to multiplication or division as that is exactly what it is doing, but consider the following two equations and their results.

```
val   = a + b * c >> 2 - d
      = 3 + 5 * 8 >> 2 - 2
      = 3 + 40 >> 0
      = 43

val   = a + b * (c >> 2) - d
      = 3 + 5 * (8 >> 2) - 2
      = 3 + 5 * 2 - 2
      = 3 + 10 - 2
      = 11
```

In general if you are having problems with your equations the problem is to do with how it is being evaluated. There are two very simple substitutions which can be done to try and correct the problem.

The first is a liberal use of parenthesis or to break up the equation into several sub equations. The second is to never loose sight of the fact that a shift operation is either a division or multiplication by a factor of two. The equation can be rewritten to do that explicitly.

Equation	Simplification
1 << 5	32
A * 192 / (1 << 3)	A * 192 / 8
	or
	A * 24
(Msb << 8) + Lsb	Msb * 256 + Lsb
(A * 46 + 22) >> 2	(A * 46 + 22) / 4

Code Blocks

The largest visual difference between the two languages is how blocks of code are defined. In python the blocks of code are intended to be visually obvious, the code at a certain indention level is grouped together.

I am not such a fan of having my formatting define my code but it may be more obvious in some situations. In this small C programming example it seems obvious that the developer wanted to calculate the average of a set of values based on the indentation.

```c
#include <stdio.h>
int main(int argc, char **argv)
{
    int I, sum = 0, avg = 0, cnt = 0;
    for (I = 11; I <= 15; I++)
        sum = sum + I;
        cnt = cnt + 1;
    avg = sum / cnt;
    printf ("sum=%d average=%d\n",sum,avg);
}
```

This program will produce an incorrect average of 65. The correct syntax for this would be to put both lines in the same block using curly braces.

```c
for (I = 11; I <= 15; I++)
{
    sum = sum + I;
    cnt = cnt + 1;
}
```

It is possible to have just about any control flow structures in C that you might have used in python, unfortunately you may need to create some of them manually in C.

Control structures

Below are some samples showing concrete examples for various control structures in python.

For Next Loop

The Python actually gives us a near error proof method of setting up a for loop. The initialization starts with zero and increments up until 5. It is possible to setup any loop increasing loop in this manner.

When reversing the loop in python another parameter is needed to change the direction from increment to decrement.

Python	C	Output
| #!/usr/bin/python<br><br>for idx in range (0,5,1):<br>    print idx | ```
#include <stdio.h>

int main(int argc, char **argv)
{
  int idx;
  for (idx=0; idx < 5; idx++)
  {
    printf("%d\n",idx);
  }
}
``` | 0<br>1<br>2<br>3<br>4 |

Regardless of the direction of your loop in C, the setup is slightly more error prone as you must explicitly do your own initialization, comparison and increment. The good news is that all these values are on one line for easy overview.

```
#include <stdio.h>

int main(int argc, char **argv)
{
    int idx;
    for (idx=4; idx >= 0; idx--)
    {
        printf("%d\n",idx);
    }
}
```

While Loop

There is almost no difference between the two languages for a simple while loop.

Python	C	Output
| #!/usr/bin/python<br><br>count=0<br>max=5<br>while count < max:<br>    print count<br>    count = count + 1; | ```
#include <stdio.h>

int main(int argc, char **argv)
{
  int count=0;
  int max=5;
  while (count < max)
  {
    printf("%d\n",count);
    count = count + 1;
  }
}
``` | 0<br>1<br>2<br>3<br>4 |

If Then Else

The "if then" statement is also nearly identical to the one you would use in C.

Python	C	Output
#!/usr/bin/python iq=125	#include <stdio.h> int main(int argc, char **argv)	really smart

```
if iq > 120:                    {
    print "really smart"          int iq=125;
else:                             if (iq > 120)
    print "average"                 printf("really smart\n");
                                  else
                                    printf("average\n");
                                }
```

If statements may be trivial and nearly identical but there is one thing that python has that C does not – which is the ability to have multiple else statements.

```
Python                          C                                   Output
#!/usr/bin/python               #include <stdio.h>                  average

iq=115                          int main(int argc, char **argv)
if iq > 120:                    {
    print "really smart"          int iq=115;
elif iq > 50:                     if (iq > 120)
    print "average"                 printf("really smart\n");
else:                             else
    print "a bit slow"            {
                                    if (iq > 50)
                                      printf("average\n");
                                    else
                                      printf("a bit slow\n");
                                  }
                                }
```

There is no limitation to the number of elif statements that you can add in python which is quite powerful. Yet even without this explicit support it is still possible to do something similar in C, it just requires a bit more effort.

Functions

Functions in python are somewhat similar to their C counterparts with the main differences being that no data types are given when creating the function.

```
Python                                          C

def <function name> ( [parameter(s)] ) :
    command1                                    <return type> <function name> ( <parameter(s)>)
    command2                                    {
    command3                                        command1
    return <return value>                            command2
                                                     command3
                                                     return <return value>
                                                }


Def mysquare (input ):                          long mysquare (long input)
    retval = input * input                      {
    return retval                                   long retval = input * input;
```

```
            return retval;
        }
```

Well there is one small difference between these two languages for creating functions. Obviously because there is no data type in python it is possible for the mysquare function to take both whole numbers and floating point numbers, where if you needed such a function in C you would need to define it for each data type.

This example function simply squares a number by multiplying it by itself and returning the value. Generally functions will be used for slightly more complex tasks than that.

Fully functional example
```c
#include <stdio.h>

int mysquare(int input)
{
        int retval = input * input;
        return retval;
}

void main(int argc, char **argv)
{
        int val = 10;
        int squared ;

        squared = mysquare(val);

        printf("value = %d\n", val);
        printf("value %d squared = %d\n", val,squared);
}
```

However, software development even with high level languages would require a lot of work if every function needed to be written by the developer him or herself. Of course there are quite a few useful libraries with standard functions for both Python and for C.

Both Python and C need to include some libraries definitions to use these functions. In Python to do that you must use the "import" keyword along with the library definition to use. In C, these definition files are called header files and tell the compiler any special values that may also be needed when using those libraries along with the function declarations. It is the inclusion of these header files that will prevent compile and linking errors when using these library functions.

Below is an example of one such warning that is generated when not including the string header file.

```
stringaddr.c:10:3: warning: incompatible implicit declaration of built-in function 'strcpy' [enabled by default]
```

Yet importing the header file for a specific library is only helpful for the compilation step when generating a program. Program generation is really a two step process. First step compiles all the source files into object code and then the second step links all the objects and libraries together to create the actual executable.

```c
#include <stdio.h>
#include <math.h>

int mysquare(int input)
{
    int retval;
    retval = input * input;
}
```

```
void main(int argc, char **argv)
{
    int val = 10;
    int squared ;
    int root;

    squared = mysquare(val);
    root = sqrt(squared);

    printf("value = %d\n", val);
    printf("value %d squared = %d\n", val,squared);
    printf("square root of %d = %d\n", squared,root);
}
```

This small code example is including the math header file in order to use the square root function "sqrt". When this is done and the math library is linked in the expected output will be displayed.

```
value = 10
value 10 squared = 100
square root of 100 = 10
```

C source code doesn't always make nice reading nor does it support strings, Python has more ways for flow control, variables need to be defined before it is used, and before running the program it has to be compiled. What is the upside to this language? Well, its really fast as the small comparison demonstrates.

Python

```python
#!/usr/bin/python

def mysquare(input):
    retval = input * input;
    return retval

def santaclaus():
    first = "santa"
    last = "claus"
    name = first + " " + last
    print name

def main():
    for idx in range(0,10):
        santaclaus();
        print "square %d" % mysquare(idx)

main()
```

C

```c
#include <stdio.h>
#include <string.h>

int mysquare(int input)
{
    int retval = input * input;
    return retval;
}

void santaclaus()
{
  char first[50];
  char last[50];
  char fullname[100];

  strcpy(first,"santa");
  strcpy(last,"claus");

  strcpy(fullname,first);
  strcat(fullname," ");
  strcat(fullname,last);

  printf("%s\n",fullname);
}
```

```
                                    int main(int argc, char **argv)
                                    {
                                     int idx;
                                     for (idx = 0; idx < 10; idx++)
                                     {
                                      santaclaus();
                                      printf("square %d\
                                    n",mysquare(idx));
                                     }
                                     return 0;
                                    }
> time ./performance.py            > time ./performance
santa claus                        santa claus

real   0m0.022s                    real   0m0.002s
user   0m0.020s                    user   0m0.000s
sys    0m0.000s                    sys    0m0.000s
```

Another reason why use C is because you are really close to the machine level. You can perform low level manipulation of values such as bit operations quite easily, but mainly it is fast.

Bit operations

This may be the part about programming that is not so beloved. Following will be some explanations of low level bit operations, for setting, testing and all around manipulation.

This is important because when dealing with the GPIO pins or communicating with devices via I2C there will be situations where it will be important for testing, setting or clearing of individual bits. Familiarity with values in either binary or hexadecimal will be key when communicating with the different devices.

Bit operations are only meaningful when looking at the number and values in binary notation. The various bit operations basically come down to turning on or off a single bit in a binary number.

People are most comfortable with numbers stored in base 10 but computers tend to store them in either binary, base 2, and developers may use hexadecimal, base 16 for representing them. Converting numbers between the different bases is actually identical to how base 10 numbers are calculated.

```
    Base 10
    6 9 2
    →           6 x 10²      = 600
    →           9 x 10¹      = 90
    →           2 x 10⁰      = 2
                             ----
                             692
```

Base 2
0 1 0 0 0 1 0 1

$\rightarrow$ 0×2^7 = 0
$\rightarrow$ 1×2^6 = 64
$\rightarrow$ 0×2^5 = 0
$\rightarrow$ 0×2^4 = 0
$\rightarrow$ 0×2^3 = 0
$\rightarrow$ 1×2^2 = 4
$\rightarrow$ 0×2^1 = 0
$\rightarrow$ 1×2^0 = 1

69

Converting into hexadecimal actually involves the same process but an easier way to do it is to convert base 2 numbers directly to hexadecimal. Simply match each nibble to the following table and substitute that value for the hex number.

Base 2 (binary)	Base 10 (decimal)	Base 16 (hex)
0000	0	0
0001	1	1
0010	2	2
0011	3	3
0100	4	4
0101	5	5
0110	6	6
0111	7	7
1000	8	8
1001	9	9
1010	10	A
1011	11	B
1100	12	C
1101	13	D
1110	14	E
1111	15	F

Converting the binary value 01000101 from our earlier example into hexadecimal we simply split our number into 0100 and 0101 and use our table. This will give us the hex value of 45h or 0x45. I prefer using the C notation of "0x" before any hex numbers. This should prevent any misunderstanding on which numbers are decimal and which are hexadecimal. I will use this convention throughout the rest of the book.

In the computer positive whole numbers are stored in binary in base 2 and with bit manipulation it is possible to change the number at the lowest possible level. Why use bits? Basically it is possible to treat an integer as a mini array of yes no values. This can be convenient and efficient to pass a single status variable around as opposed to up to 32 different boolean status values or an array of status values.

The individual bits can be set, tested or cleared one at a time or en mass. Binary number are base two the only two values are zero and one. Depending on the architecture the bits are numbered zero through n from either right to left or left to right.

Little endian

7	6	5	4	3	2	1	0

Big endian

| 0 | 1 | 2 | 3 | 4 | 5 | 6 | 7 |

The Raspberry Pi is a little endian machine[51]. The endianness is actually not very important when writing software if these values are only stored in memory, this is not the case if the values are stored in a database or file and then loaded on another machine with a different endianness or when software is ported from a different machine with a different architecture.

Bit manipulations involves the actual value which is evaluated with a mask to perform a certain function.

Logical Or
Setting an individual bit is done with a "logical or" operation. A logical or is performed to ensure that the same bits from the mask are set. It doesn't matter if the bit is already set, when either of the bits are set to one then the resulting bit will also be set to one. The following example shows the outcome from the four possible cases.

Logical OR

| 0 | 0 | 0 | 0 | 1 | 1 | 0 | 0 | ← Number

| 0 | 0 | 0 | 0 | 0 | 1 | 1 | 0 | ← "OR" Mask

| 0 | 0 | 0 | 0 | 1 | 1 | 1 | 0 | ← Result

```
int number = 0x0C;
int mask = 0x06;
printf("%02x", number | mask)
> 0E
```

The result of this operation has bits one through three are set to one after performing a logical or.

Logical And
Testing that a bit is set is done with a logical and. The mask is used to test if any or all of the bits in the mask are set in the number being tested. The resulting number will a binary number with the bits set that matched in both the number and the mask.

Logical AND

| 1 | 1 | 0 | 0 | 1 | 1 | 0 | 0 | ← Number

| 1 | 0 | 0 | 0 | 0 | 1 | 1 | 0 |

---

51 Well, the Raspberry Pi can actually be big endian as well. Technically it is a BI-Endian machine. This tends not to be problem when writing software for the pi from scratch and not copying the binary data to other machines that have a different endianness. http://www.raspberrypi.org/phpBB3/viewtopic.php?f=2&t=531

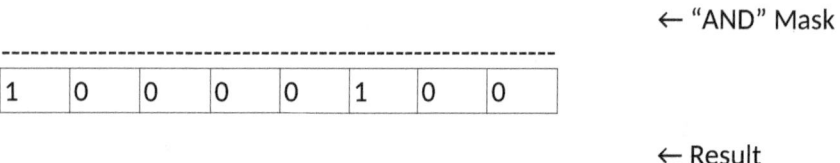

← "AND" Mask

← Result

The resulting number is non-zero if any of the bits were set in the number to be tested as well as the mask. This information may or may not be enough information for the developer. When testing multiple bits at the same time if any of the bits are set then the result will be non-zero. In order to determine if all of the bits tested are set the resulting number must match the same value as the mask.

```
int value = 0x72;
int mask = 0x36;

if (value & mask)
{
    if (value & mask == mask)
        printf("it looks like all of our bits are set");
    else
        printf("it looks like some of our bits are set");
}
```

Using the "logical and" operator will allow us to test if individual bits are set, but it will also allow us to clear individual bits from a status field.

Simply create a mask that has all bits set to one except for the bit or bits that need to be cleared. Perform a "logical and" between the status value and the mask and the resulting value is the status value with all fields from the mask cleared.

Logical AND

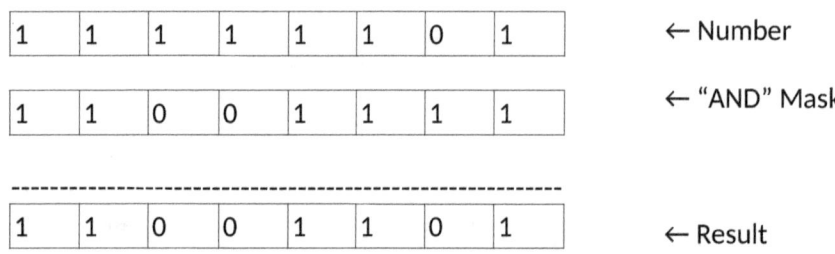

← Number

← "AND" Mask

← Result

It is easy enough to create such a mask to clear a field or fields but depending on the number of different status values and combinations would make it tedious and error prone to setup all these masks.

```
int value = 0x6E;          // decimal 110
int clearmask = 0xBF;      // decimal 191
value = value & clearmask;
// value will equal 0x2E   // decimal 46
```

Fortunately, there is an easier way.

Complement

The complement operation can be used to toggle all bits in a value to their opposite value. This provides a convenient way to take a mask and toggle the values so it can be used to clear that bit or bits.

Compliment

| 0 | 0 | 1 | 0 | 0 | 0 | 0 | 0 |

| 1 | 1 | 0 | 1 | 1 | 1 | 1 | 1 |

This makes it quite convenient to take a individual flag or and programmatically convert it to a mask which can be used to clear that individual value.

```
// example 1
int value = 0x6E;                   // 110
int testmask = 0x40;                // 64
value = value & ~testmask;
// value will equal 0x2E            // 46

// example 2
int value = 0x6E;                   // 110
int mask1 = 0x40;                   // 64
int mask2 = 0x80;                   // 128
int mask3 = 0x03;                   // 3
value = value & ~(mask1 |
            mask2 |
            mask3);
// value will equal 0x2C            // 44
```

Shift Left and Shift Right

The shift operator will shift all bits in a binary value to the left or right as many places as desired.

| 0 | 0 | 0 | 0 | 0 | 0 | 0 | 1 | ← Before shift

| 0 | 0 | 0 | 0 | 0 | 0 | 1 | 0 | ← After shift

A left shift has the same effect as multiplying the value by two, while a right shift has the same effect as doing a division by two. The shift operator can move the bits multiple places to left or right.

This operator simply takes a value along with the number of places to shift. If you simply want to multiply or divide by 2 the shift operation will do everything you want and do it quickly.

```
int variable = 0x17;
int shiftamount = 2;
printf("%02x", variable << shiftamount));
> 5C
```

This operator can also be used to create masks to help with either setting or clearing of individual bits. Simply take a one and shift it into the proper location.

Depending on what you find to be more intuitive, you can use this operation along with a macros to create different masks.

```
#include <stdio.h>
```

```
#define BIT0   0
#define BIT1   1
#define BIT2   2

int clear_nth_bit (int input, int bit)
{
    printf("%d %02x\n",bit, 1 << bit);
    return input & ~(1 << bit);
}
int main(int argc, char **argv)
{
    printf("%02x\n",clear_nth_bit(255,BIT0));
    printf("%02x\n",clear_nth_bit(255,BIT1));
    printf("%02x\n",clear_nth_bit(255,BIT2));
}
```

This example uses the actual bit numbers to create the mask, but it is also possible to simply have the different bit mask defined directly. One possible difference is that this method will require calculations to create the mask while macro that has the actual value can simply be used.

```
#define BIT0        0x01
#define BIT1        0x02
#define BIT2        0x04
```

Exclusive Or

The final operator is the exclusive or. This operation will compare each bit and when one bit is a zero and the other is a one, then the resulting bit will be a one. Conversely when both bits are either zero or are one then the result is zero. Basically said, this operator can be used to toggle the value of a bit to its opposite value.

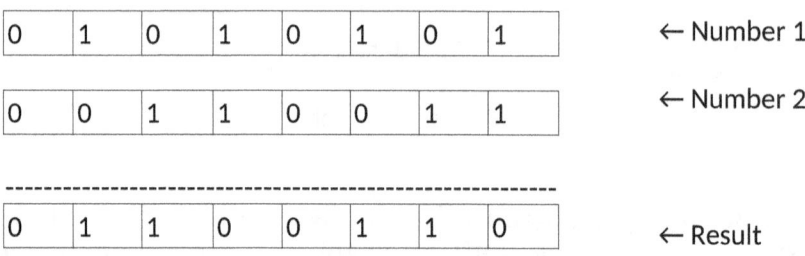

It may be somewhat less clear in which situation you wish to use this particular operation, but a very simple case would be to toggle the value of some status value with each iteration.

```
#include <stdio.h>
int main(int argc, char **argv)
{
    int on_off_indicator;
    int loopcounter;
    int mask = 0x01;

    on_off_indicator = 0x01;
    for (loopcounter = 0; loopcounter < 10; loopcounter++)
    {
        // do something here

        // change state
        on_off_indicator = on_off_indicator ^ mask;
        printf("%d\n",on_off_indicator);
```

```
        // do more things after this
    }
}
```

This is a vivid example of how to use this operator for toggling a bit from its existing state to the other state.

There are other uses for the "exclusive or" but this bit operator is not as commonly used as the ones described thus far.

Other operators
There are other operations that are quite similar to the ones that we have already discussed. These operator are actually more like shortcuts as they help to condense down some of the operations. They don't need a lot of special description as most of them are fairly obvious.

    Variable = Variable + 10;   →   Variable += 10;
    Variable2 = Variable2 + 1;   →   Variable2++;
    Variable3 = Variable3 * 4;   →   Variable3 *= 4;

The goal is not to elaborate all of the operators that are available in C or Python but rather to discuss one additional operation - modulo division. Modulo division differs from regular division as it returns the remainder of an integer division not the quotient.

Concept	Symbol	Example
Modulo Division	%	A = 15
		B = 10
		print A % B
		> 5
		C = 10
		D = 8
		print C % D
		> 2

Pointers
In C there are many different types of variables, from 8 bit characters, 32 bit integers and low and high precision floating point numbers. There is one more type of variable which is a bit different from the rest – the pointer variable. This variable actually doesn't hold the value itself, but rather it holds the value of the memory address which contains the variable's value.

It is no exaggeration that the ability to directly access memory from C is very powerful. A high level programming language with low level power, but pointers in C is much akin to Pandora's box. Pointers may seem very simple but they have the power to wreak complete havoc.

Pointers provide the tools for creating other power types of data structure such as linked lists and binary trees just to name a few. Linked lists and binary trees are truly interesting data structures. When adding additional items to these structures you allocate the memory directly and simply add the pointer to that memory to your list or tree. Pointer is probably the trickiest topic in C development. Rather than to elaborate all of the powerful things you can do with pointers we need to look at a single aspect of pointers.

When calling functions it is possible to pass in variables as input and it is possible for the functions to return values when exiting. In order to manipulate either variables or structures from within a function they would need to be a global variable or would need to be passed in as pointer. In this way, the function is not manipulating the variable but directly accessing the memory location of that variable.

The following program is an example showing how a pointer can be used. The function "simple_function" changes the input variable which demonstrates that the value is changed but the change only affects the local copy of the variable not the one from the calling function.

```c
#include <stdio.h>

void simple_function(int inputonly)
{
    inputonly = 10;
    printf("in simple_function %d\n",inputonly);
}

void powerful_function(int *input_output)
{
    *input_output = 88;
}

int main()
{
    int integer = 999;

    printf("before simple_function\n");
    printf("integer: %d\n",integer);
    simple_function(integer);

    printf("after simple_function\n");
    printf("integer: %d\n",integer);

    printf("\n");

    printf("before powerful_function\n");
    integer = 999;
    printf("integer: %d\n",integer);
    powerful_function(&integer);

    printf("after powerful_function\n");
    printf("integer: %d\n",integer);
    return 0;
}
```

The function "powerful_function" is passed the address of our variable, e.g a pointer to an integer, so when the value is changed the actual variable in the calling function is changed.

There is one small difference when using pointers, it is only possible to assign a pointer address to a pointer variable. When a value needs to be assigned to the memory address pointed to by the pointer, you need to use the address. This is done by "dereferencing" the pointer, e.g using the variable storage is being used.

```c
void powerful_function(int *input_output)
{
    *input_output = 88;
}
```

There is nothing wrong with using pointers but they are a very error prone feature of C as can be seen by the following few example lines.

```
int *ptr;                    pointer to an integer
ptr = 0x3234;                pointer is now pointing to memory address 3234h
*ptr = 42;                   memory address 0x3234 contains the value 42
```

Warning: You do not directly assign memory addresses yourself.

C++ Programming refresher

C++ programming can be fun but despite the support for C++ on the Arduino, the Arduino is a platform with a fairly meager set of hardware resources. Full fledged C++ programs would take up too many of those precious resources.

13 MAKEFILES

As previously mentioned, Python is an interpreted language and does not require any special steps in order to run a script – but this is not the case with C. The C programming language is a compiled language and you must run the compiler, gcc, against the different source files and link them together into a executable or a library in order to somehow use the result.

The make program is a developer tool to control the creation of executables from their source code. The "Makefile" is the default name of the file containing the dependencies of the files and the rules for their generation. The number and type of rules that can exist in a Makefile are arbitrary and can range anywhere from running test, installing software or even cleaning up intermediate files. You are only limited by your imagination.

One of the tasks we want to perform is the compilation of our code to object files and the linking of our object files and libraries to create our executable.

```
<rulename>:        <dependancies>
        <command 1>
        <command 2>
        ...
        <command n>

stoplight.o:       stoplight.c stoplight.h
        @echo building $@
        gcc -I. -c stoplight.c -o stoplight.o
```

This rule will run when either the stoplight.o file doesn't exist or if the source or header file is newer than the object file which would indicate that the source code has changed. The first rule will echo "building stoplight.o". The $@ symbol is replaced with the name of the rule. The line for this rule runs the compiler with telling it to use include files from the current working directory (-I.) and do not perform the linking step (-c) and the resulting object file should have the following name (-o stoplight.o)

```
stoplight:  stoplight.o
        @echo building $@
        gcc stoplight.o -lm -o stoplight
```

This rule although virtually the same as the previous one is an example of how to link in a required library. The only difference is the -lm option which will link in the "libm.a" library when creating the executable.

This may seem like a lot of effort for compiling a single file and actually it is. The reason is this is a very simple project. The rules would not change very much even if we added a few more source files.

```
mainstoplight:      stoplight.o street.o trafficrules.o
        @echo building $@
        gcc stoplight.o trafficrules.o street.o -lm -lstreetsmarts -o $@
```

It is actually easy to add a few more objects to our program, although I haven't listed the rules for compiling them, they would be virtually identical to the stoplight.o rule. This is really easy to have a rule for each object when there are one or two extra source files, but when there are dozens or hundreds of files it is considerably easier to put them all together into a library.

A library is a number of object files packed into a single archive which can be later linked into programs. The first requirement is to compile the source files into object files. This can be done by setting up a rule for each individual object file or using some more generic setup to limit the number of explicit rules that need to be defined.

In order to make things more flexible, make allows us to setup some variables with some of the values we wish to use in our rules. These variables can be used directly in any of our rules or commands much in the same way as in the shell.

```
VAR1=something
echo $(VAR1)
```

It is possible to even use these variables while defining new variables.

```
#   for library
LIBRARY=libmylcd.a
LIBSRC=mylcd.c mylcdsupport.c
LIBOBJ=$(LIBSRC:.c=.o)
```

The most important variable in this example is the LIBOBJ variable. The LIBOBJ variable is being defined to the same value as the LIBSRC variable but all the .c file extensions are replaced by .o file extensions. Thus we do not need to have multiple variables each maintained by hand with a list of source and objects that must be kept in sync.

```
.c.o:
    @echo "compling $@"
    @gcc -c $(CFLAGS) $< -o $@
```

The rule ".c.o" is a generic rule for compiling source code. When an object file is required that doesn't have its own specific rule defined, this rule will compile the source into an object file.

```
$(LIBRARY):   $(LIBOBJ)
    @echo "creating library $@"
    @ar rcs $(LIBRARY) $(LIBOBJ)
    @ar -s $(LIBRARY)
```

The command for creating the library is the ar. This tool will create a library with a given name, in our case defined by the LIBRARY variable, from our list of objects.

```
LIBOBJ=mylcd.o mylcdsupport.o
```

Compiling code into libraries or executables is what make was intended for but make can be setup to perform other tasks as well.

```
.phony: clean
```

```
clean:
    @echo cleaning up build env
    @rm $(LIBOBJ) $(LIBRARY) $(OBJ) $(PGM)
```

Right before the clean rule is another rule called ".phony". This rule is required for special targets. The presence of this rule guarantees that the dependency will be run regardless of modification date or even if that object exists.

Another rule example that actual doesn't create an object when it is run is the install and uninstall rules. These rules simply perform commands that could also be run from the command prompt, yet have been placed into the makefile for convenience and so they are always executed in an identical manner.

With the install logic in the make file it is possible to install a new library by simply running make for the install rule. This will need to be done as a privilaged user or using the sudo command. Once the new library has been installed simply add it to the list of libraries in our makefile and it will be linked in just the same as any standard library.

```
.phony: install
install:    $(LIBRARY)
    @echo installing library
    @install -m 0755 -d $(DESTDIR)$(PREFIX)/lib
    @install -m 0755 $(LIBRARY) $(DESTDIR)$(PREFIX)/lib
    @install -m 0755 mylcd.h $(DESTDIR)$(PREFIX)/include
    @install -m 0755 mylcdsupport.h $(DESTDIR)$(PREFIX)/include

.phony: uninstall
uninstall:
    @echo un-installing library
    @rm -f $(DESTDIR)$(PREFIX)/include/mylcd.h
    @rm -f $(DESTDIR)$(PREFIX)/include/mylcdsupport.h
    @rm  $(DESTDIR)$(PREFIX)/lib/$(LIBRARY)
```

Last but not least is the definition of a few dependencies so make can verify that the object built is as current as the header file it depends on.

```
#
# some dependancies
#
mylcd.o:    mylcd.h
```

When the make command is run it will verify that all objects are up to date building any that are not. It is common to have an "all" or "world" rule defined first which actually is only a list of the main targets. When the make program is executed without any parameters the default rule, which is the first rule defined in the Makefile, will be executed. It is important to know that the first rule declared will be executed by default.

Simple Makefile
This simple Makefile can be used for single file programs. Simply change the variable PGM to the name of the file with your code.

```
PGM=bmp085

#
# nothing below this needs to be changed
#

SRC=$(PGM).c
OBJ=$(SRC:.c=.o)
CFLAGS=-Wall
```

```
INCDIR=-I.
LIBS=-lbcm2835 -lm
LIBDIR=

all:    $(PGM)

$(PGM): $(OBJ)
        @echo building $(PGM)
        gcc $(INCDIR) $(OBJ) $(LIBS) $(LIBDIR) -o $(PGM)

.c.o:
        @echo "compling $@"
        @gcc -c $(CFLAGS) $< -o $@

clean:
        @echo cleaning up build env
        @rm -f $(OBJ) $(PGM)

.phony: run
run:    all
        sudo ./$(PGM)

#dependancies
$(PGM).o: $(PGM).h
```

This very simple Makefile is including two libraries, the math library and the bcm2835 library. These are only here to demonstrate where and how to add new libraries, if they are not used by your program they are not needed.

Library Makefile

This Makefile is for library development. You can have a small list of source files for the library and as well as a small single file application that can be used for unit testing the library.

```
#
# variables
#

#  for compiling
CFLAGS=-Wall
INCDIR=-I.

#  for linking
LIBDIR=-L.
LIBS=-lmylcd -lbcm2835

#  for application
PGM=myapp
SRC=$(PGM).c
OBJ=$(SRC:.c=.o)

#  for library
LIBRARY=libmylcd.a
LIBSRC=mylcd.c mylcdsupport.c
LIBOBJ=$(LIBSRC:.c=.o)

#  for installing
DESTDIR=/usr
PREFIX=/local

#
# rules
#

all:    app library
```

```
library: $(LIBRARY)

app:   $(PGM)

.c.o:
    @echo "compling $@"
    @gcc -c $(CFLAGS) $< -o $@

$(LIBRARY):   $(LIBOBJ)
    @echo "creating library $@"
    @ar rcs $(LIBRARY) $(LIBOBJ)
    @ranlib $(LIBRARY)

$(PGM): $(OBJ) $(LIBRARY)
    @echo "building $@"
    @gcc $(OBJ) $(LIBDIR) $(LIBS) -o $@

.phony: run
run:   all
    sudo ./$(PGM)

.phony: clean
clean:
    @echo cleaning up build env
    @rm $(LIBOBJ) $(LIBRARY) $(OBJ) $(PGM)

.phony: install
install:    $(LIBRARY)
    @echo installing library
    @install -m 0755 -d $(DESTDIR)$(PREFIX)/lib
    @install -m 0755 $(LIBRARY) $(DESTDIR)$(PREFIX)/lib
    @install -m 0755 mylcd.h $(DESTDIR)$(PREFIX)/include
    @install -m 0755 mylcdsupport.h $(DESTDIR)$(PREFIX)/include

.phony: uninstall
uninstall:
    @echo un-installing library
    @rm -f $(DESTDIR)$(PREFIX)/include/mylcd.h
    @rm -f $(DESTDIR)$(PREFIX)/include/mylcdsupport.h
    @rm  $(DESTDIR)$(PREFIX)/lib/$(LIBRARY)

#
# some dependancies
#
mylcd.o:    mylcd.h
```

Compilation Problems

Not all languages are created equal in more ways than one. Some may be better suited for a particular task, while others may be easier to develop quick solutions or perhaps some may be more efficient with the machine resources.

Some problems that can occur with C or any compiled language tend to manifest themselves only during the compiling or linking steps. I have listed a few problems that can occur as well as the corrective steps.

Undefined reference

This output is informing us that the functions log and exp which are used in the nthRoot function cannot be located when trying to link the final executable.

```
gcc nthroot.c -o nthroot
/tmp/ccwanQHj.o: In function `nthRoot':
nthroot.c:(.text+0x1e): undefined reference to `log'
nthroot.c:(.text+0x32): undefined reference to `exp'
collect2: error: ld returned 1 exit status
make: *** [bad] Error 1
```

This problem actually has nothing to do with our code at all.

Nthroot.c
```c
#include <stdio.h>
#include <stdlib.h>
#include <math.h>

double nthRoot(double value, int n)
{
    double retval;
    retval = exp(log(value)/(float)n);
    return retval;
}

int main(int argc, char **argv)
{
    float val = atof(argv[1]);
    int n = atoi(argv[2]);
    double nth;

    nth = nthRoot(val,n);
    printf("the %sth root of %s = %0.5lf\n",argv[2],argv[1],nth);
    return 0;
}
```

Looking at the message the problem is not that log or exp are not defined as far as the compile step goes, it is just when it tries to link the program to those methods.

```
gcc nthroot.c -o nthroot
```

The compile step is simply executing the compiler against our code but these methods are in the math library and it isn't being linked in. We simply need to explicitly link it in.

```
gcc nthroot.c -lm -o nthroot
```

The option -l is used for linking in a library. The naming convention for libraries is lib<name>.a and as we want the math library we needed to add -lm when compiling this code.

Multiple Definition

This problem is also not one of compilation but rather one of linking. In the linking step we see that the function writemessage is defined multiple times.

```
gcc device1.c device2.c myapp.c -o myapp
/tmp/cc5AOdsr.o: In function `writemessage':
device2.c:(.text+0x0): multiple definition of `writemessage'
/tmp/ccrIrqHJ.o:device1.c:(.text+0x0): first defined here
collect2: error: ld returned 1 exit status
```

This seems pretty obvious for this example as all the code will fit on a single piece of paper. We can see that both the device1.c and device2.c contain a function writemessage.

device1.c
```c
#include <stdio.h>
#include "device1.h"

double writemessage(char *msg)
{
    printf("Error: %s\n",msg);
}
```

device2.c
```c
#include <stdio.h>
#include "device2.h"

double writemessage(char *msg, double v)
{
    printf("%s %lf\n",msg,v);
}

double calc_value(double a, double b)
{
    return a * b;
}
```

myapp.c
```c
#include <stdio.h>
int main(int argc, char **argv)
{
    writemessage("this that and the other thing");
    double v = calc_value(10,20l);

    return 0;
}
```

This small example is pretty obvious, we have two functions with exactly the same name. When the linker tries to make sense of it simply points out our inconsistency. The solution is to rename all of our functions so there are no name collisions.

If we were creating a library of functions, the easiest way would be to prefix each function with the name of the device or function.

Conflicting Types

This problem is actually no different from the previous problem, but in this case it is because of a global variable not because of a function. Well, the problem is slightly different as this is a problem during the compilation of our code.

```
gcc device1.c device2.c myapp.c -o myapp
In file included from device1.c:3:0:
device2.h:3:5: error: conflicting types for 'a'
In file included from device1.c:2:0:
device1.h:2:8: note: previous declaration of 'a' was here
In file included from device2.c:3:0:
device1.h:2:8: error: conflicting types for 'a'
In file included from device2.c:2:0:
device2.h:3:5: note: previous declaration of 'a' was here
In file included from myapp.c:4:0:
device2.h:3:5: error: conflicting types for 'a'
In file included from myapp.c:3:0:
device1.h:2:8: note: previous declaration of 'a' was here
```

In the two header files the variable "a" is defined which is actually not very good style, but even worse, it is defined as two different data types.

device1.h
```c
double device1_writemessage(char *msg);
double a;
```

device2.h
```c
double device2_writemessage(char *msg);
double device2_calc_value(double a, double b);
int a;
```

There are a couple of things that can be done to make this a cleaner solution.

device1.h

```
#ifndef DEVICE1_H
#define DEVICE1_H

double device1_writemessage(char *msg);

// global variables are not the best style
extern double device1_a;
#endif
```

device2.h

```
#ifndef DEVICE2_H
#define DEVICE2_H

double device2_writemessage(char *msg);
double device2_calc_value(double a, double b);

// global variables are not the best style
extern int device2_a;
#endif
```

As C has no functionality for creating separate name spaces, the global variables and functions have all been prefixed with the name of their "device" or "function". The variables have been moved into their respective source files but are defined in the header file using the extern keyword. It is important that the actual variable is defined in one of the source files otherwise there will be a linker error. The linker error would be similar to the following.

```
myapp.c:(.text+0x17): undefined reference to `device1_a'
```

The extern keyword tells the compiler that this is just the definition of something and that there will be an actual declaration of the function or variable elsewhere. This is why we should have an extern keyword before the definition of the variables and functions in our header file.

It is not required that you declare your variables in this manner nor include guard1 statements. If the include file is only included once this will not be a problem, nor will multiple inclusions be a problem either – if the variables or functions are the same.

It is good style to include guards in your header files to ensure that they are included only one time.

```
#ifndef DEVICE2_H
#define DEVICE2_H

#endif
```

This code will be used by the pre-compiler to ensure that the contents of the include file is included only one time. This will save time during the compilation step by not including files multiple times.

These are just a small handful of possible errors that can occur when trying to build C programs, but the error messages are actually reasonably informative of to what the problem is.

14 KICAD

Kicad is a computer aided design program which rather than a single monolithic program is a collection of different programs that provide different views on the electronic circuit that is being designed.

The Kicad program is the launching point for all the related programs from creating schematics to the printed circuit board. When creating a new project, you will be prompted if a directory should be created. It should as this is both best practice as well as a convenient way to keep all the associated files together without mixing them up with any other projects.

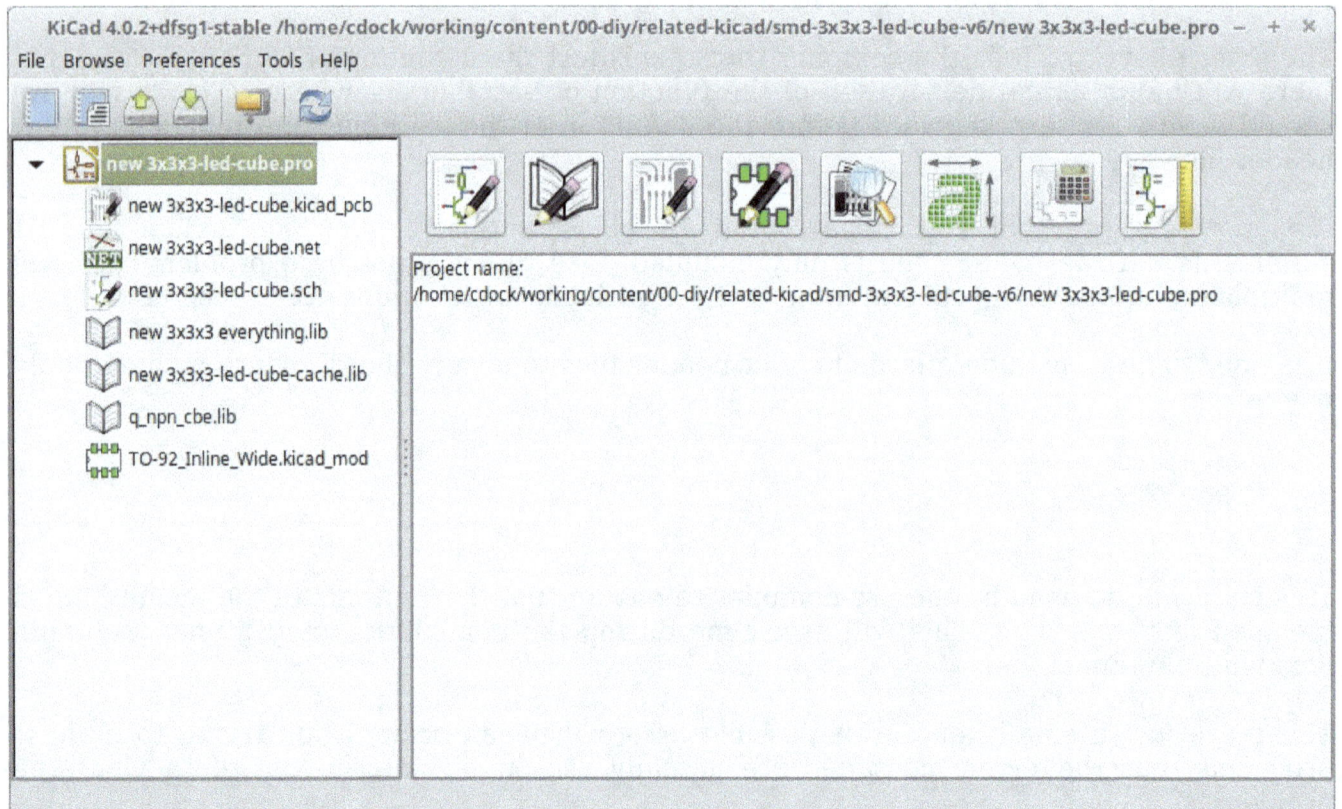

Illustration 114: Kicad main dialog

Schematic editor

The schematic editor allows for the creation of a schematic which is just a diagram that shows how all of the parts are connected together at an electrical level.

Footprint editor

The footprint editor which can be used to create footprints for new components or components not in the Kicad libraries. It is thus possible to add your own unique footprint or the footprint for a part that is not yet known to Kicad.

PCB editor

The printed circuit board editor allows for the design of the layout of the PCB. It takes the schematic and allows the user to control everything that involved in the creation of the PCB. Starting with the placement of the components and layout of the traces connecting them together.

Schematic Editor
The thing to remember about the schematic is that this is not a picture of how the components are soldered to the PCB but a description of how all parts interact with each other.

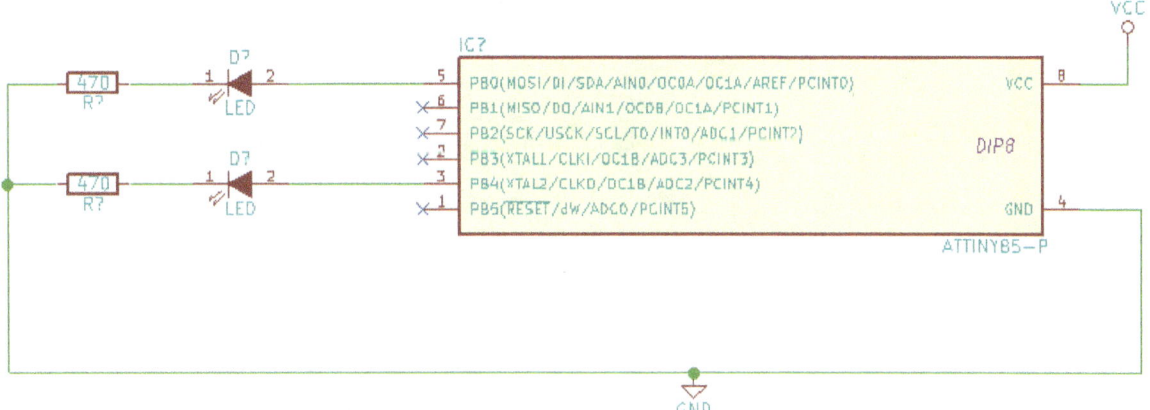

Illustration 115: Schematic with all wires connected

This simple diagram is to show an example schematic where not all parts are directly connected. In this case the connection simply ends with a label. Kicad knows that all tracks that end in a label with the same name are actually connected in the same net on the PCB.

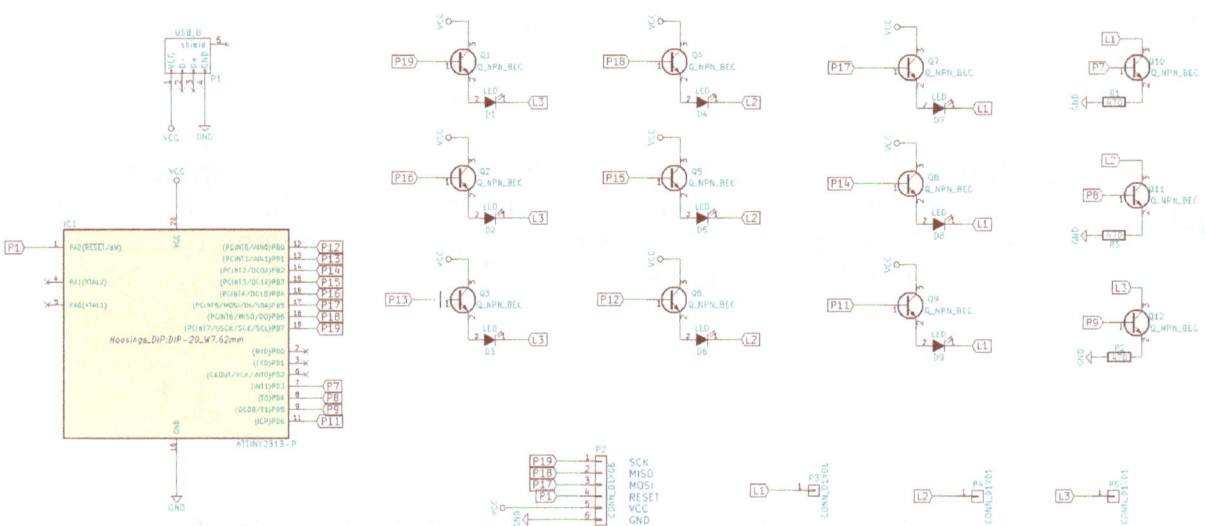

Illustration 116: Schematic with labels showing relationships

Kicad CAD software is actually just different views of the same object. This may be the schematic or elements on the schematic. Not only does this tool allow you to work on the different parts but there is one more important view of the schematic – the PCB viewer.

The goal of using Kicad or some other CAD software package is to create a printed circuit board. You can keep track of which parts are being used by the circuit board with any tool from a piece of paper to an excel spreadsheet but that does create the situation where data can get lost.

The good news is that it is possible to store component information along with each of the components in the schematic itself.

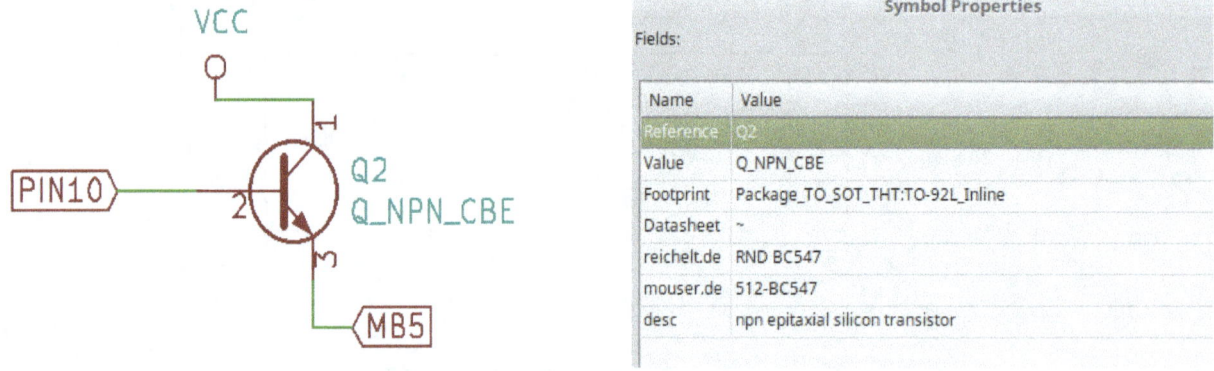

Pressing the letter "e" will bring up a list of meta data fields associated with that components. Some of these values are created by Kicad and contains values that cannot be changed directly such as the reference name. The reference name can be changed by

Kicad if more parts are added to the schematic and the annotate option is selected. Yet it is possible to use this data area for other information that should also not get lost.

In this example, I have added two different vendors Reichelt and Mouser as well as the actual supplier part number from those web sites. The reason I have two different suppliers is in case I have problems sourcing the parts. Should that situation occur I am already prepared with a backup supplier.

This is actually very easy to enter this data when adding a part to the schematic but it is highly suggested that this be done when placing the part. The reason for this is because it is not unusual for a circuit board to have multiple identical components and the easiest way is to duplicate a part that is already placed. All additional meta data fields will also be duplicated when a part is duplicated in the schematic editor. Thus, ensuring that the part contains any other important meta data will reduce work in the long run. If this is not done, then the amount of work needed later is multiplied by the number of similar components when this data is updated at the end.

Other important steps
As previously mentioned Kicad isn't a single monolithic program but rather a collection of tools that are all used towards the goal.

Assigning footprint
Footprint is just another word for the size, shape and the arrangement of pins of an electronic component. The footprint editor is the link between electrical connections, the schematic, and the physical PCB.

There are a lot of different electronic components but more importantly most of these components can exist in different shapes and sizes. If the component is a simple resistor or capacitor the size and shape of it is not important for the schematic. The size and shape are extremely important when trying to design the layout of the printed circuit board.

One example of this is the Atmega 328P micro-controller. This micro-controller comes in a surface mount package but it also comes in a through hole package. The differences couldn't be greater. The through hole requires holes in the board in order to solder it while the surface mount device demands that no holes exist but rather a nice flat spot exists for each pin. Not only that but the shapes and sizes of these two components differ as well.

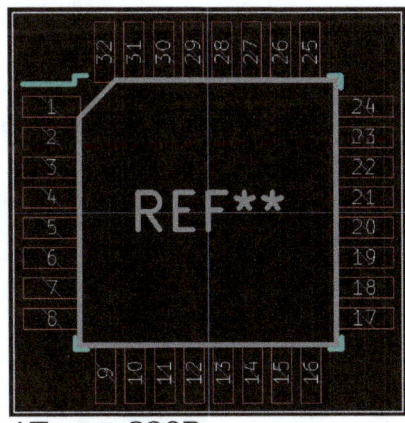

ATmega328P
32-pin thin quad flat pack

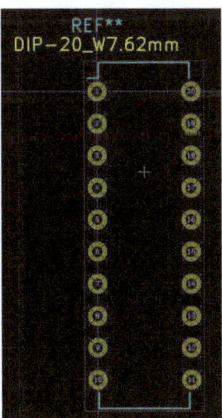

ATmega328P
28-pin narrow DIP

When placing the individual components on the schematic they have no physical description. In order for the component to get this physical description it needs to be assigned. This assignment process is selecting the physical layout so it can be used for each of the components.

The assignment process is simply a three-pane dialog. The left is a list of all the footprint libraries, the middle pane is the parts from your schematic and the right panel is the actual footprints.

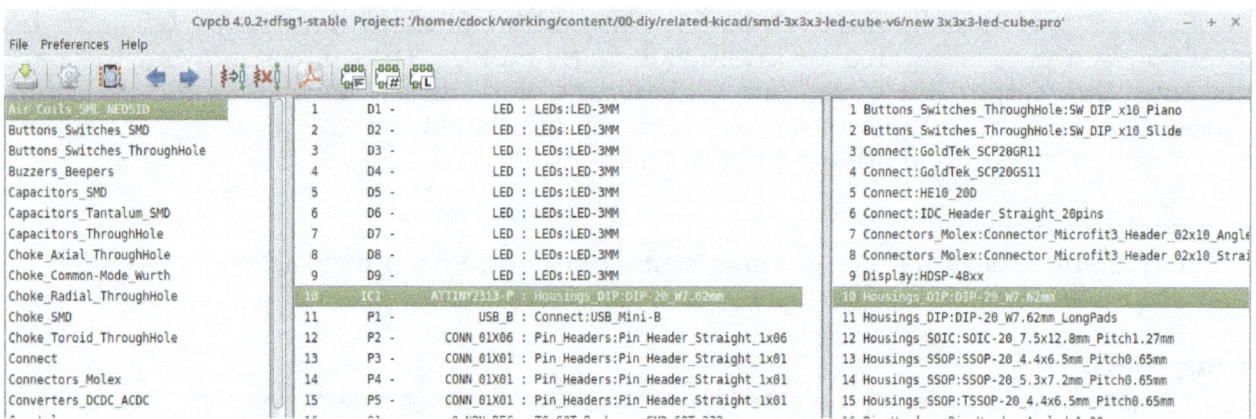

Illustration 117: Assign footprints dialog

The footprint editor is just a mapping of the component (middle) with the actual design (right). The pane on the left is a list of the libraries which contain the various layouts.

Annotating
When creating a schematic, you place components on the diagram. When placing multiple identical components they are still generic with each one looking exactly the same and they would be indistinguishable to Kicad when attempting to create a PCB layout. This makes the process of any sensible verification impossible. The solution to this is that a process is to ensure that each component gets a unique id which will allow the CAD software to notify when problems are discovered.

Extracting the net list
The Kicad suite is a collection of different programs. The schematic editor is a pure view of the electronics of a solution while the PCB editor is a view of the final product. There are some requirements of the PCB editor such as footprint that need to be fulfilled in order to have enough information to create the layout. This extra information is eventually stored with the schematic but this additional information comes from the annotating and footprint selection steps. Thus an additional step is required to copy this information from the schematic so it can be used by the PCB editor. This step is called extract netlist in the Kicad schematic editor. When the schematic is finished along with the selection of footprints and the annotations then this data is exported. This data will then be imported into the PCB editor so it is aware of what the footprints of the components should be.

Shortcuts
Kicad can perform most if not all operations in multiple ways. One way is to use a mouse to click through menus or on one of the toolbars. A second method is to memorize the keyboard shortcuts for these same operations.

Not all shortcuts are very intuitive but once learned they do speed up the creation and manipulation of schematics and PCB boards.

Schematic editor - keyboard shortcut reference

Task	Key	Description
Adding a component	a	This brings up a combo box of all components known to Kicad. Simply select which component should be added.
Adding text	Double click	This will open up a dialog so the text can be entered. The text can then be placed on the schematic, moved or even copied just like any other component.
Copying a component	c	This will copy the component pointed to by the cursor. This will include any component properties that have been defined.
Connecting a wire	w	This will connect a wire to the component lead pointed to by the cursor which can then be dragged to the next component and then connected.
Deleting wire or component	Del key	This will delete a component, wire or wire segment that the cursor is hover over.
Dragging a component	g	Drags a component while not severing any of the connections.
Editing properties	e	This will open the component or label dialog so the properties can be edited.
Moving a component	m	Picks up a component, severing all connections, so it can be dropped elsewhere.
Placing a global label	Ctrl-Alt-H	Click on the sheet where this should be created and then this will open up a dialog box for the net name. This will be created as a global label. This will logically connect all components connected to a label with this name from all sheets. The physical connections must be made when doing the PCB layout.
Placing a net	Shift-L	This will open a dialog box for the net name. This will be created as a local label and all components connected to this label on the sheet will be connected. This is different than a global label.
Zoom in	F1	Zoom out while keeping focus where the cursor is.
Zoom out	F2	Zoom in at the point where the cursor is.

PCB Editor

Creation of the printed circuit board layout is both the most complicated and the most rewarding part of the process.

When designing a board there are several different rules that should be considered but there is one that must be followed. That rule is that the unrelated traces cannot intersect with each other. Careful part placement and clever routing of the traces can reduce the risk of an intersection but there are situations where some component is in the way. The main solution for this situation is to have multiple layers.

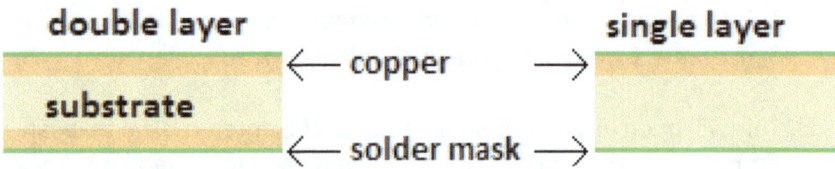

Illustration 118: Side view of pcb

Single and double-sided boards are the simplest cases. It is possible for very complex projects to have up to 8 layers, but it was possible back in 2001 to create boards with 100 layers.

Multiple layers allow you to connect through hole parts from either side of the printed circuit board. This is one method that can be used to ensure that none of the traces intersect when they should not. This is a neat way to add flexibility to your board design but not all electronic components are through hole components.

The ability to miniaturize components has reached some amazing levels. The number of transistors in a modern CPU today has exploded to 592 million today in the Intel I7 compared to 275 thousand in the Intel 80386 in 1985.

The only difficulty of using this level of miniaturization is the connections to the outside world. The input and output pins can only be made so small. To make these individual components smaller required a change in format from through hole to surface mount technology.

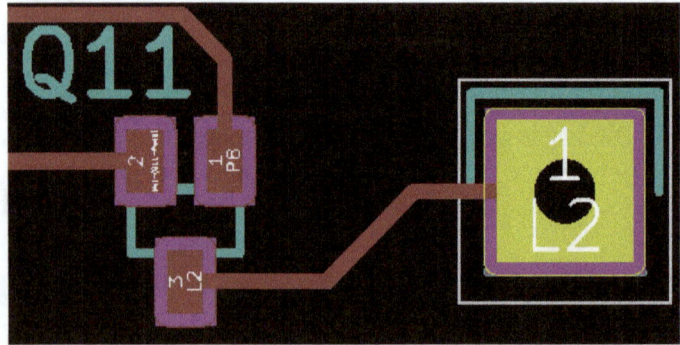

Illustration 119: Surface mount part connecting to a through hole

Surface mount, like the name suggests, are mounted to pads on a surface of the board and do not have a hole through the PCB. The advantages that SMD parts bring is the reduced size but also allow automated PCB assembly.

Yet even with smaller SMD components designs can still encounter situations where traces would intersect with something that it shouldn't. This problem can also be eliminated using via's.

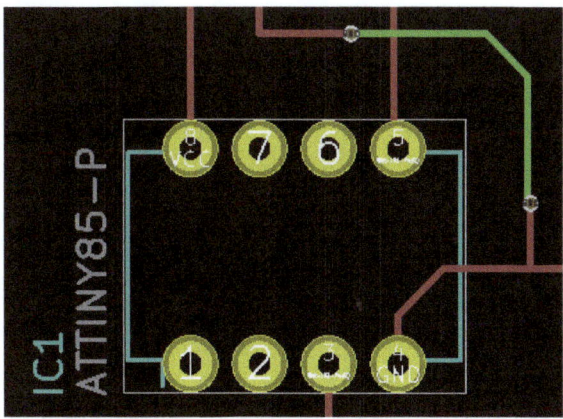

Illustration 120: Example of a via

A via is essentially just a tunnel through the circuit board that connects a trace from one side with a trace on another layer. This example shows a trace that needs to be rerouted so a via is used. We can see that this is on a different layer because of the different color of the trace.

Unlike the schematic editor this is the actual layout that will be used by the manufacturing company to create the board. Thus all connections that were described in the schematic as connected to each other must connected, there is no cheating the placement complexity in the same way it is possible with labels in the schematic editor.

Guidance
When designing a circuit board it may be important that the traces are at 45 degree angles not 90 degree angles to prevent logic signal feedback. This type of problem may occur when designing a board that does a lot of high speed communication.

Traces need to be sized correctly. The amount of current that can be transmitted is limited by the width of the individual traces. This is especially important for traces that are powering the rest of the board.

Power should laid out in a star configuration versus a daisy chained ideally with the traces supplying the power being of equal length.

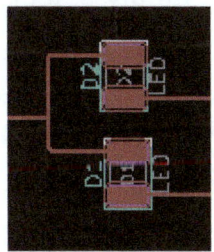
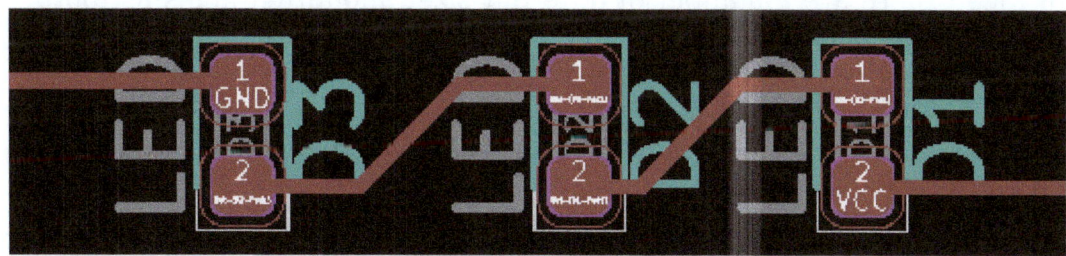

Illustration 121: Star configuration *Illustration 122: Daisy chained configuration*

The reason for this is so each section or component gets an equal share of the power.

Otherwise the components closer to the power supply will take its power and there will be a voltage drop resulting in less power available for parts farther from the power supply.

Parts should be placed in with the same polarity orientation when necessary. This is to assist the person soldering the parts to the board. If the orientation of the parts are not consistent it is more likely to solder a part incorrectly.

Make sure to leave space between your traces. This is to help reduce the possibility of short circuits. The same is true for pads. Pads also shouldn't be too close. Even if the manufacturing process goes perfect it is possible to create a short circuit when you are soldering the parts to the board.

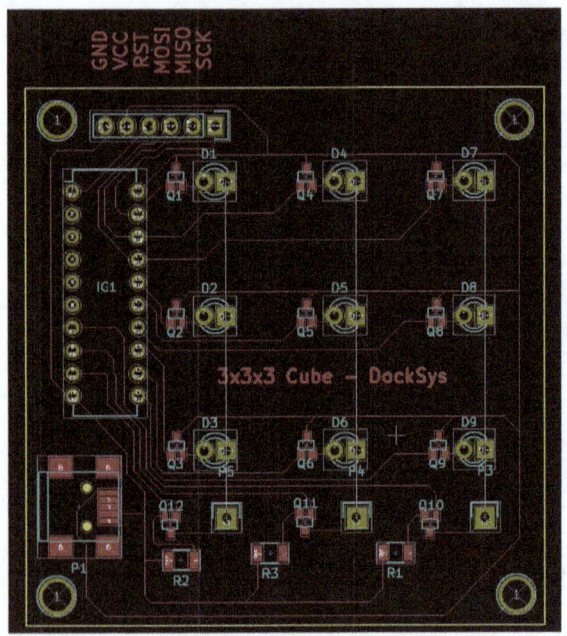

All components in this editor are displayed using their selected footprint. Despite the obvious visual differences this editor is quite similar to the schematic editor.

One of the major differences between the PCB layout editor and the schematic editor is that you do not normally add new components when editing the PCB, you simply import the list of components from the schematic, move them to their final location and connect them to each other.

The components, when imported, do not actually contain the circuit traces but instead show the relationships (ratsnest) between the various components. The suggested relationship connections disappears as each component is connected.

One of the final steps for a printed circuit board is to add some mounting holes. These would be an exception to my earlier statement that no parts get added to the circuit board in this editor.

To add a mounting hole you either select add footprints from the toolbar or from the menu. In Kicad there are a number of different sized mounting holes as footprints. Simply select which one is most suitable.

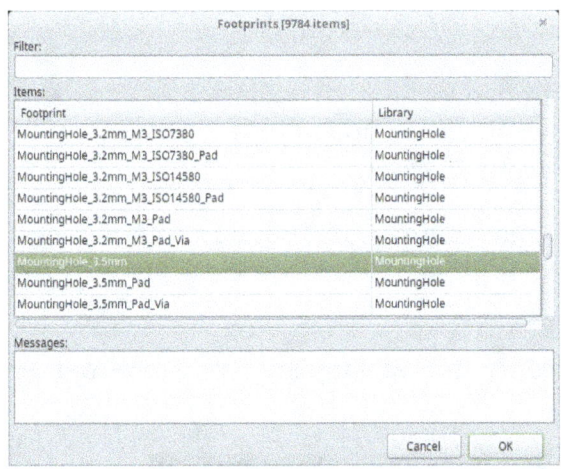
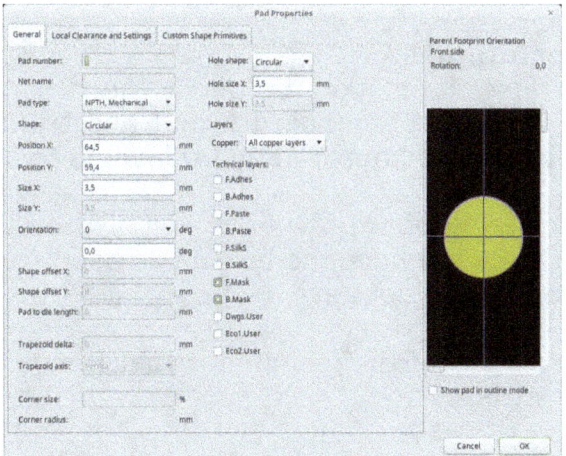

It is also possible to change the actual diameter or shape of the hole. In Kicad it is then possible to select any of the footprints that exist but these footprints are special. They are special because they do not exist in the circuit and thus it is not possible to take a trace from your layout and to connect it to a newly placed footprint.

The final step would be the size and shape of the circuit board. This layers is used to communicate with the manufacturer what the final board shape should look like.

PCB editor - keyboard shortcut reference

Task	Key	Description
Adding text	Shift+control+T	This menu choice will open up a dialog so the text can be entered. The text can then be placed on the schematic, moved or even copied just like any other component.
Connecting a wire	w	This will connect a wire to the component lead pointed to by the cursor which can then be dragged to the next component(s).
Deleting wire or component	Del key	This will delete a component, wire or wire segment.
Dragging a component	g	Drags a component while not severing any of the connections.
Editing properties	e	This will open the component or label dialog so the properties can be edited. This is a different set of properties from those in the schematic editor. This set of properties describes the component or portion of the component (e.g one of the pads).
Moving a component	m	Picks up a component, severing all connections, so it can be dropped elsewhere.
Zoom in	F1	Zoom out while keeping focus where the cursor is.
Zoom out	F2	Zoom in at the point where the cursor is.
Add footprint	O	Brings up a dialog allowing you to select a footprint. This is most helpful when adding a mounting hole.

Other techniques
Just like any other sport or talent the best way to improve it is to practice. The best practice is to take any board layout and try to improve it. It is not actually necessary to bring any of those improved board layouts to a manufacturer.

Creating a new board layout is just like software development. It can be changed and modified daily but once the board has been manufactured older pcb versions, especially for small projects, do not have a lot of value.

Below are some of the techniques that I have discovered while attempting to create better circuit boards.

The only rule when designing printed circuit boards is that the traces cannot cross. This can lead to traces winding around the components or even around the entire edge of the board. Two overlooked options are underneath the chip or even between the pins when the space is wide enough.

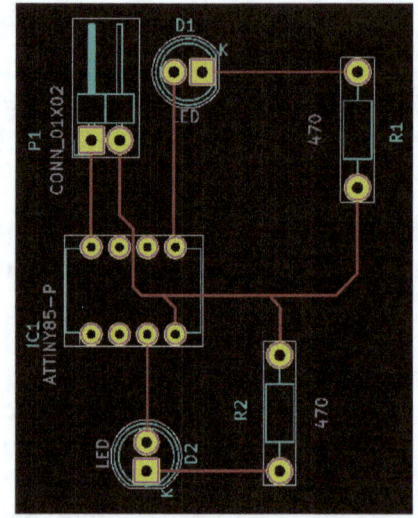

Traces underneath a chip is a convenient method of getting all traces connected without the use of via's. This same technique can be used with any components. Traces can go underneath through hole resistors or between any parts with legs or pins that have space.

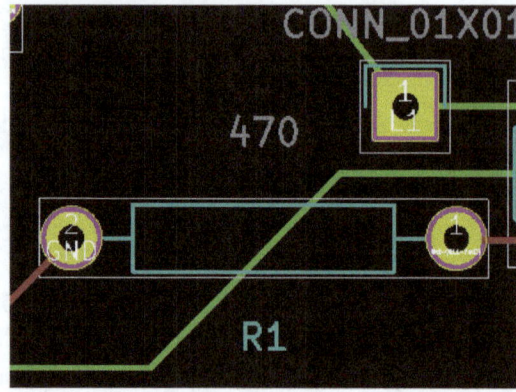

It is possible to combine traces underneath other components with a via. A via is just a tunnel between two different layers of the board. This might be from the front side of the board to the back or perhaps to a layer that is between the top and bottom layer.

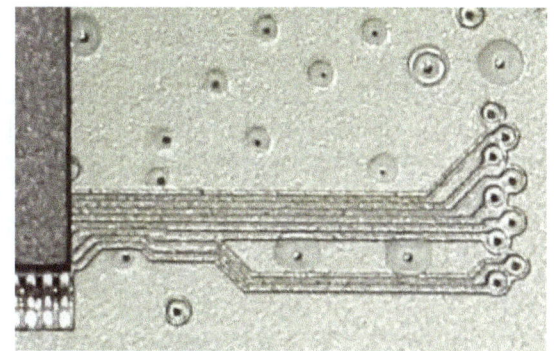

Using multiple layers is an obvious way to connect different parts. SMD components are small which does allow a high density but can complicate trace placement.

If any through hole parts are used it is possible to connect traces on a different layer. This is essentially the same technique as a via.

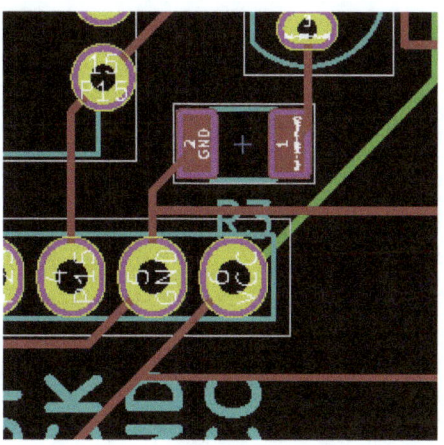

Generating Gerber files

The act of generating the set of Gerber files will vary from CAD tool to CAD tool. However as many of these tools will probably be offering the same data in similar ways I will show how to export the Gerber files for Kicad. The process in Kicad is actually fairly literal as the option for creating these files is called "Plot". This option brings up a dialog which either contains data that must be selected or a summary of other global values.

Kicad 4.07

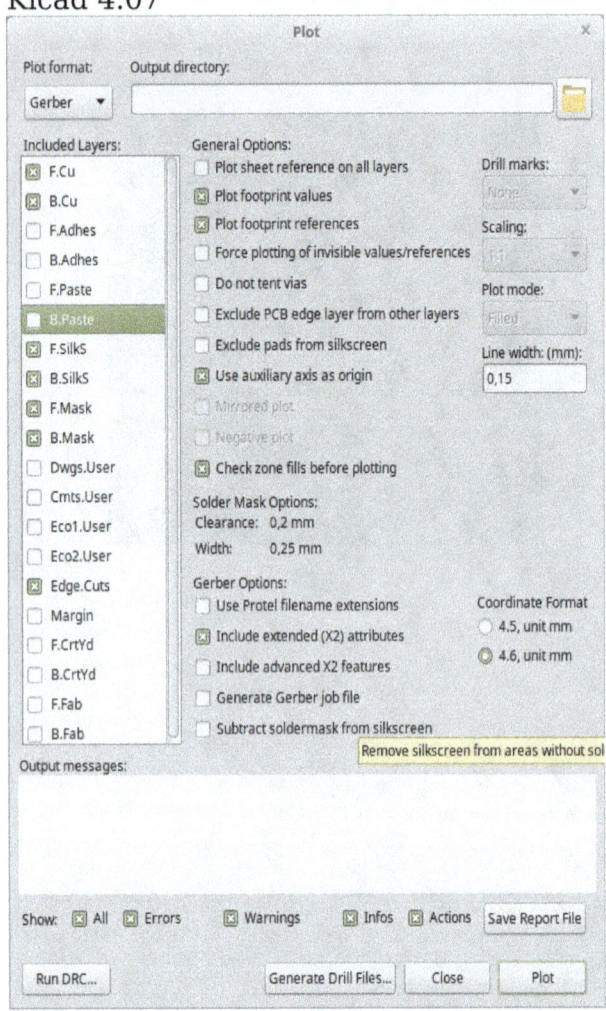

F.CU	Top copper
B.CU	Bottom copper
F.SilkS	Top silkscreen
B.SilkS	Bottom silkscreen
F.Mask	Top solder mask
B.Mask	Bottom Solder mask
Edge cuts	Size and Shape of board

One of the files that is required but is not actually a Gerber file is the drill file. The drill file is actually an Excellon file format.

Kicad 4.07

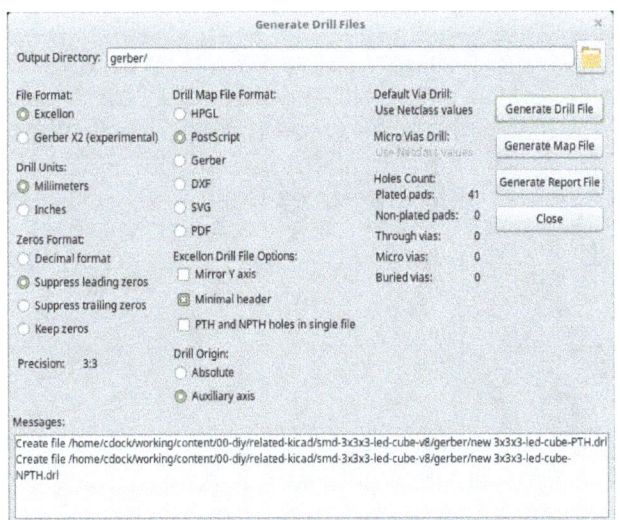

Most of these values are actually the defaults. Those small changes I did make were the result of what the expectations of the manufacturer.

Kicad 5.1

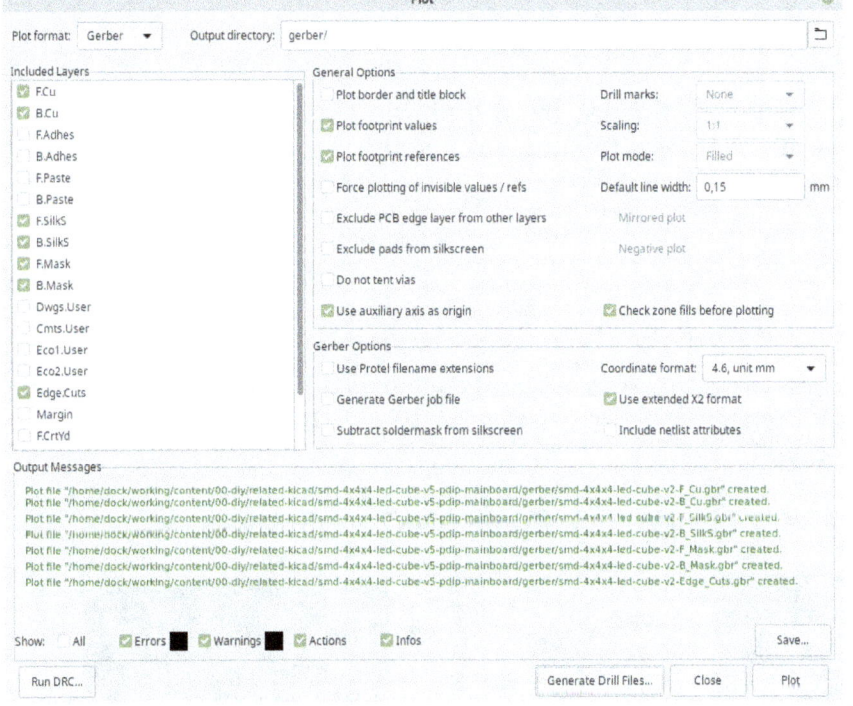

F.CU	Top copper
B.CU	Bottom copper
F.SilkS	Top silkscreen
B.SilkS	Bottom silkscreen
F.Mask	Top solder mask
B.Mask	Bottom Solder mask
Edge cuts	Size and Shape of board

Kicad 5.1

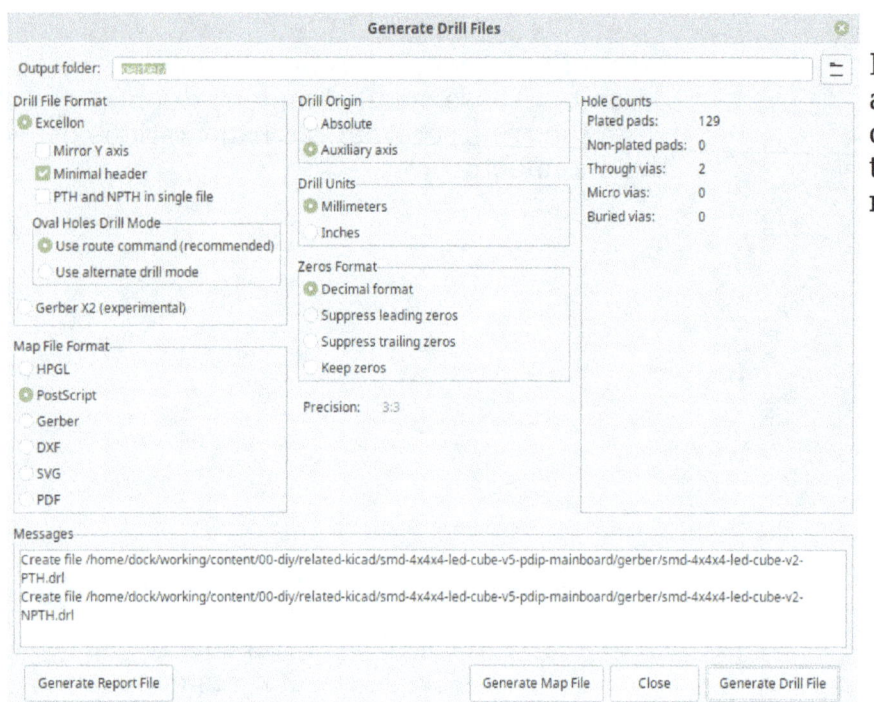

Most of these values are actually the defaults. The changes were the result of what the requirements of the PCB manufacturer I use.

Because Gerber files are not the proprietary format from a single CAD software there is a lot of support for them. There are a number of different open source tools available as well as online tools.

http://www.gerber-viewer.com
http://circuitpeople.com
http://mayhewlabs.com/3dpcb

Because of the importance of Gerber files there is actually a very good chance that your CAD software already has a viewer for them. It is also quite likely that the manufacturer also has an online Gerber viewer as well.

I wanted to see just how similar my board looked in different Gerber viewers.

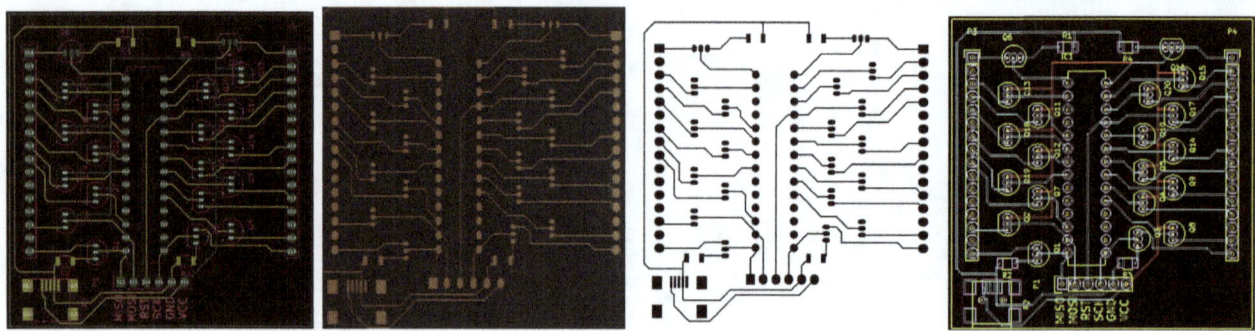

This is not a completely fair comparison as I only used online viewers and not all viewers allowed you to show all layers at the same time.

If your eyes are keen enough you may see things that look suspicious either while examining your PCB layout or perhaps viewing the Gerber files. I discovered my problem when comparing my export of the Gerber files with the recommendations of my manufacturer.

It may not be completely obvious but if you look at the layout you can see that the solder mask clearance (purple line) is really thin. This is confirmed when looking at the system settings. The solder mask clearance is 0.05mm.

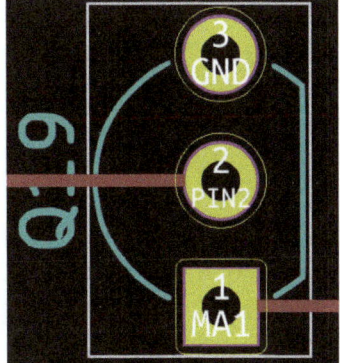
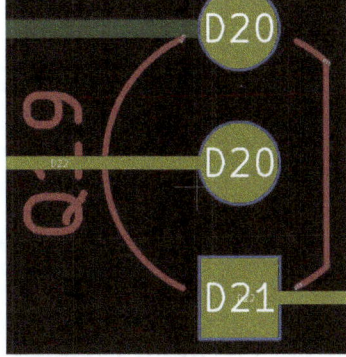
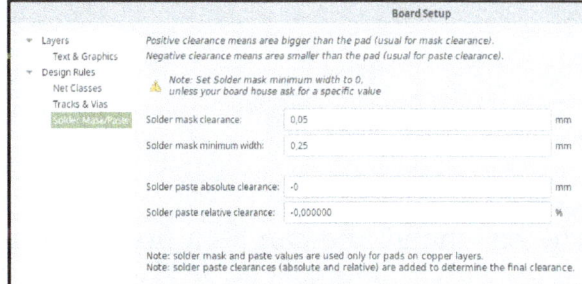

PCB Layout   Gerber viewer   System defaults

You can see that there is a big difference by changing the solder mask clearance value.

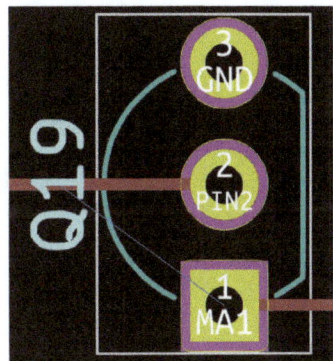
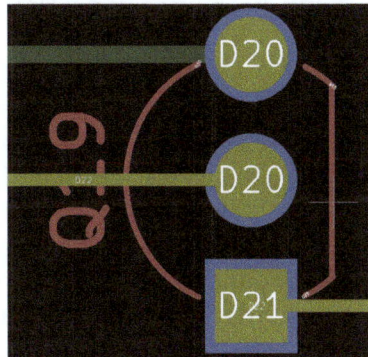

PCB Layout   Gerber viewer

In version 5 of Kicad the solder mask clearance is located in the board setup (File → Board Setup).

15 LINUX COMMANDS

The command line can be a pretty boring place, however, a lot of the power configuration and scripting happens here. This is even more the case if you end up using your Raspberry Pi as headless computer that drives some custom appliance or device. Whether or not if you boot directly to the console you will always be able to open a shell which is where the magic happens.

When connecting to the Raspberry Pi via secure shell, you will also have a command line. Thus to be comfortable there are some common commands that you will need to use in order to move around.

The disk is arranged as a hierarchy of directories each of which can contain one or many files. Most Linux command names have a tenancy to be very small but meaningful once you understand what the command name means.

List files

The list files command will display the files that exist in a certain directory. The "ls" command in its basic form will simply show the files and directories that are in the current working directory. This command, like most Linux commands, takes additional parameters to alter the behavior of the command.

```
Common options
    -l                    detailed (long) listing format
    -t                    sort by date and time, newest first
    -r                    reverse the order when sorting
    -i                    show inode number of the file
```

It is not only possible but common to use more than one of these options.

    ls -l -r [path]

        - or -

    ls -lr [path]

When no path is given the directory listing will be of the current working directory.

Copy files
Files may need to be copied to another media or sometimes it is easier to copy a file and edit it to produce a script with similar functionality. The "cp" command can copy a single file but it can also copy an entire sub-directory.

Common options
- -p   preserve time stamps and modes when copying.
- -t   sort by date and time, newest first
- -r   -copy directories recursively

cp -p <source file> <destination file>

cp -r <source directory> <destination directory>

Change directory
The directory structure is hierarchical and rather than typing long directory and filenames it is usually easier to change to different directories. The "cd" command does exactly what the name implies.

You can either change to a different directory in either absolute or relative terms. This command actually doesn't take any other parameters other than the new directory. If no directory is given you will change to your home directory.

cd [path]

It is possible to move to a directory higher in the hierarchy by using the special name "..". Thus just the change directory with the path ".." moves to the parent directory.

Displaying file contents
Sometimes you want or need to see the contents of a text file. Especially when this file is small the easiest way to view this information is to have it printed to the shell. The "cat" command will display the output of the file to the terminal. Obviously this is usually more meaningful when the file contains text as opposed to binary data but interesting things can be done with binary data using cat.

cat [path]<filename>

Paginating output
Displaying the contents of a file using the "cat" command may display more information than will easily fit on the screen. The "more" command will display one screen full of text and then pause. Pressing the spacebar will advance to the next screen while pressing the enter key will advance by one line.

more [path]<filename>

The more command will displays single pages of text from a file to the terminal but this command can be used in a second manner. Text that is displayed to the terminal can be sent through more using the pipe (|) command. The output of some other command is then attached as input to the more command. Data that is piped through more will behave exactly the same as if the input came from a file.

cat [path]<filename> | more

Print current directory
The file system can have a lot of sub-directories and sometimes it is easy to get lost. The command "pwd" is short for print working directory. This command will simply display the directory that you are currently in. It may seem obvious which directory you are in however, this is quite useful in scripts.

    pwd

Make directory
A good way of organizing data is by creating sub-directories or entire trees of sub-directories. The "mkdir" directory command like the change directory command takes either absolute or relative directory paths. You can only make a single sub-directory at a time and only in directories that exist.

    Common options
-p	This option will create all necessary sub-directories in a path
-v	Print out each directory as it is created

It is also possible to use both of these options together

    mkdir -p -v [path]<directory>

      - or -

    mkdir -pv [path]<directory>

Moving files
When organizing your data files or directories may need to be moved to other locations. The "mv" command will move either files or directories to another location irrespective of the type. No rename command exists, thus this same command can be used to rename a file.

    Common options
-f	Do not prompt before overwriting
-i	Prompt before overwriting
-v	Display what is being done
-u	Move only if source is newer than destination file, or destination file is missing.

    mv -f [path]<source> [path]<directory>

Remove
The opposite of creating sub-directories is the ability to delete sub-directories. The command for removing directories is called "rmdir". This command will remove an empty directory. However, there is a more powerful method for removing files.

    Common options
-f	Do not prompt
-i	Prompt before each removal
-r	Remove directories and their content recursively
-v	Display what is being done

The "rm" command is for removing files but can also be used to remove directories empty or not. This is done by using both the -f and the -r command. This will recursively remove the files and will not prompt before doing so.

    rm -rf <path>

These commands are probably all you will need to traverse the directory structure as well as other general options.

File and Directory permissions
Windows supports file permissions although they may not be used very often by most users. Depending on what version of windows, windows server or file system you are using will depend on the list of actual permissions.

Unix-like operating systems have a much simpler permission structure. Each file or directory is owned by who creates it and the group of that user. In addition to these pieces of ownership information there are three separate sets of permissions. These will grant read, write or execute permission to the owner, group and all(e.g public access).

```
$ ls -l
drwxrwxr-x  8 cdock cdock        4096 Mar 26 20:28 pictures
-rw-rw-r--  1 cdock cdock    99425002 Mar 29 20:39 Arduino.odt
-rwxr-x---  1 cdock users        5002 Mar 29 20:39 converter.sh
```

When doing a directory listing using the long listing format (e.g ls -l) you also see the permissions that have been granted.

    File permissions
    d r w x r w x r w x
    →   d = directory
      →   owner read permissions
        →   owner write permissions
          →   owner execute permissions
            →   group read permissions
              →   group write permissions
                →   group execute permissions
                  →   world read permissions
                    →   world write permissions
                      →   world execute permissions

Note: When the associated value (e.g r) is replaced with a dash(-) then this means that this value has not been granted for that person or group.

In the above file example, the converter.sh script can be modified and run by the owner, can be run by the users in the same group, and all others cannot view the contents of this file but also cannot run this script.

It is possible to change the owner of a file, group of a file as well as the individual permissions that a file possess.

    chown   changes owner of a file.

	chown <user> <file>
chgrp	changes group ownership of a file.
	chown <group> <file>
chmod	changes permissions of a file.
	chmod <permissions> <file>

The permissions can be added in a fairly friendly manner. It is possible to set or reset each individual permission.

chmod a+r	file is now readable by all.
chmod a-r	cancels the ability for all to read the file.
chmod g+rw	group and read and write the file.
chmod u+rwx	user can read, write and execute.

16 ADVANCED LINUX COMMANDS

Input, Output and Errors
One of the well organized things of Unix is that there are three standard channels for a program to communicate with the user - these are input, output and error. This allows command line programs to communicate with the outside world. A program can read data from the input channel to be processed and can write out its results to the output channel. Should some error occur while running then it can be written to the error channel.

When a program is running in a bash shell the output and the error output appear to be mixed together but they are not and it is possible to separate these from each other. The errors can be saved in a logfile while the calculated output could be put into its own file.

From the bash shell, it is possible to do this separation with the use of greater than and less than symbols.

    myprogram < input.txt
    myprogram > output.txt
    myprogram > output.txt 2> error.txt

This particular syntax from the command prompt connects to the input, output or error channels directly with the files given.

Pipes
The concept of pipes is actually no more than the connection of the input and output channels. It was a really clever discovery back in the day when it was decided that you can connect the output from one program to the input of another. This allows you to create chains of commands but one of the strengths of this approach is that the first program must not finish all of its processing before the next program in the chain can begin processing. The data that is passed between processes is buffered so it is possible that both programs are actually running concurrently.

This configuration makes it possible for each link in the chain to winnow down the data. The final result is usually a smaller amount is in some way processed so it can be immediately used (e.g sorted).

The syntax for connecting these input and output channels between programs on the command line is the pipe symbol.

    ls -l | more

This command will do a long file listing of the file in the current directory and if that list is greater than will fit on a single screen it will be paused by the more command until the user presses a key.

It may not be obvious from this description but there are no real limitations to the number of links in this "chain". Linux can support a single pipe as easily as eight. Despite the efficiency of doing this, it should be done with care. Not because Linux has a problem but the next person who needs to modify the script needs to be able to understand the logic.

    cat `ls -1 | head -3 | tail -1` | more

Command breakdown
Before the cat command can be run the portion between the back ticks will be evaluated.

ls -1	This will send only file names of the files in the current directory to the output stream.
head -3	This will show the top three names from the standard input stream.
tail -1	This will take the last item from the standard input stream

Once this small chain of commands have been run, the resulting output will be a single filename. The cat command will then use this filename and pipe the results through more. Thus the contents of this file will displayed to the console. If the output is longer than will fit on the console then it will be paginated by the more command.

Tee – viewing and logging output
Command line programs can produce lots of output either as regular output or as errors. Using the various redirection symbols or pipe symbol it is possible to capture the output to a file but this can make following the process somewhat more difficult.

Difficult is not impossible. The tail command will display the last few lines from a text file. This command can be run to not only display those few lines but to display all output that subsequently ends up in the file. Yet, there is an even easier way to follow this output.

The "tee" command reads the input from standard input and displays all output to the screen while at the same time it will be saved to an output file.

Common options
    -a   Append to given file do not overwrite

    tee [-a] <output file>

Awk – a standard scripting language
There are many different programming languages which can be used to write a new command line program or application to calculate or process data. The number of languages to do this are too many to cover in detail but the top few would be C, C++ and Java.

There are quite a few different situations where writing a heavy duty solution is either too time consuming or resource intensive. The good news is that there are just as many lighter scripting languages available such as python, perl or ruby. Any of these are suitable but if you are interested in just quickly parsing out a few values there is yet another Linux command which can do this – AWK.

It is actually possible to write programs in AWK but it is quite often used for quickly parsing a value from a line of input. It is especially useful in situations where the data to be processed is in columns with a common separator.

This program has been covered in various books, my favorite is Sed & AWK from O'Reilly books. Rather than to recreate hundreds of pages of information documented elsewhere, I will demonstrate a common use for shell scripts. AWK is used for parsing out individual values from a string.

Common options
- -f <program file>   Location of AWK program script
- -F <separator>   Used to separate each field, default is space

awk [-F <separator>] [-f awk script] <input file>
awk [-F <separator>] [-f awk script]

AWK can be used either to process data from a datafile or can be used to process data from standard input.

$ ls -l
drwxr-xr-x 14 dock dock   4096 Sep 23 22:48 daily
drwxr-xr-x 21 dock dock 12288 Oct 13 00:50 Downloads
drwxr-xr-x 18 dock dock 12288 Oct 13 17:24 Documents

$ ls -l | awk '{ print $9}'
daily
Downloads
Documents

Awk is a proper scripting language so it is not in anyway limited to parsing directory listings.

$ echo "chris;dock;blue" | awk -F \; ' { print $3 }'
blue

wc – word count
The word count program, wc , can count characters, words or lines. It can be used to calculate the number of items to be processed. This command much like many other Linux commands can be used both for processing data from a datafile or from the standard input stream.

Common options
- -c   Number of characters
- -l   Number of lines
- -w   Number of words

wc [-c] [-w] [-l] <datafile>
wc [-c] [-w] [-l]

The wc command can be used for calculating any of these values and more but it will calculate all three by default if no specific values has been decided via command line options then all three will be displayed.

```
$ ls -l | wc -l
35

$ ls -l | wc
35          310         2184
```

nohup - no hangup
When starting a program from a shell it will continue to run until it completes or until the shell ends. Depending on the situation this may be ok but there are circumstances where the program needs to run to completion. In these cases it is possible to use the nohup program to run the program until it completes regardless if the calling shell terminates first.

nohup COMMAND [args]

ps - process list
Every program that is running on Linux or even on windows is some individual process. Most of the time you don't need to look at which process is running. This information is available in windows with the task manager and in Linux this information is displayed with the ps command.

Like most Linux commands this will output the information to the console. The ps command can display more or less information about the processes depending on what information is required.

```
Common options
    -e              Show all processes
    -f              Full format list
    -af             show current processes and their dependencies

$ ps
  PID TTY        TIME CMD
 2113 pts/0   00:00:00 bash
 2170 pts/0   00:00:00 oosplash
 2187 pts/0   00:05:17 soffice.bin
 4025 pts/0   00:00:00 ps


$ ps -f
UID      PID  PPID  C STIME TTY        TIME CMD
dock    2113  2105  0 15:08 pts/0  00:00:00 bash
dock    2170  2113  0 15:08 pts/0  00:00:00 /usr/lib/libreoffice/program/oosplash arduino5.odt
dock    2187  2170  4 15:08 pts/0  00:05:17 /usr/lib/libreoffice/program/soffice.bin arduino5.odt –splash-pipe=5
dock    4026  2113  0 17:06 pts/0  00:00:00 ps -f


$ ps af
  PID TTY      STAT   TIME COMMAND
```

```
 2113 pts/0    Ss     0:00 bash
 2170 pts/0    Sl     0:00  \_ /usr/lib/libreoffice/program/oosplash arduino5.odt
 2187 pts/0    Sl     5:17  |   \_ /usr/lib/libreoffice/program/soffice.bin arduino5.odt --
splash-pipe=5
 4027 pts/0    R+     0:00  \_ ps af
  1230 tty7        Ssl+       1:15 /usr/lib/xorg/Xorg -core :0 -seat seat0 -auth
/var/run/lightdm/root/:0 -nolisten tcp vt7 -novtswitc
 1262 tty1     Ss+    0:00 /sbin/agetty -o -p -- \u --noclear tty1 linux
```

kill - stopping a process

Knowing which processes are currently running is not normally at all interesting. These process numbers are mainly used if a process stops running correctly and cannot be stopped in any other way.

Common options

--<signal number> <pid>	Send signal number to given process
-s <signame> <pid>	Send signal number to given process
-l	List all signals

Most common used signals

sig name	sig #	description
SIGKILL	9	unblockable kill
SIGTERM	15	termination signal
SIGINT	2	keyboard interrupt

The most interesting item in the list of signals is SIGKILL. This is interesting because it is listed as unblockable kill implies that the rest of the signals can be blocked. Actually, programs can listen for signals and decide how to react to them. The only signal that they cannot capture is SIGKILL.

tar

Tar is a very common Linux archiving tool. Similar to the Windows winzip command it is used for taking files or entire directories and storing it in a single file.

Common options

-l	List number of characters
-v	Verbose
-c	Create an archive
-f <file>	Following filename should be used for archive
-x	Extract files from archive

tar [options] <file>

The tar command has a tremendous amount of functionality. If it is being used for either creating backups or extracting source code so a program can be built then the listed options will be sufficient.

gzip

The gzip utility despite almost sharing a name with the popular zip behaves slightly different. The gzip utility will only compress a single file into a gzip archive. This might not sound very useful but the Unix philosophy is for each program to do one thing well. Thus the gzip program is often used with the tar program which will combine one or many files into a single archive which can then be compressed by gzip. An interesting technical note is that gzip and zip have similar internals so very often any program that can deal with zip files can typically unpack gzip files.

    Common options
        -9   Maximum compression, will take longer to run
        -d   Decompress

    gzip [OPTIONS] <filename>

bc – precision calculator

It is possible to do some small calculations just in the bash shell without any additional programs. If the problem requires more than simple arithmetic or requires higher precision then using the Linux command line program bc comes to the rescue.

    Common options
        -l   Uses standard math library

    bc [OPTIONS] < "<filename>"

Instead of being limited to integers arithmetic the bc command supports fractional values.

    $ echo "7 / 5" | bc -l
    1.40000000

This can be for single line calculations but it can do much more. It is also possible to limit the number of important decimal places for your calculations.

    $ echo "scale=3; 7/5" | bc -l
    1.400

Because the input for the bc command is coming via standard input it is possible to create variables in your shell or shell script and echo those values as part of your equation.

    $ A=7
    $ B=5
    $ echo "scale=2; $A / $B" | bc -l
    1.40

The precision of calculations is controlled by the addition of the "scale" instruction. This level of flexibility and essentially integration with the command line is more than enough reason to make this a invaluable resource for shell scripts. Yet bc is even more powerful than that.

If you embed the bc command in a slightly different manner into your script you can actually define your own functions. This doesn't necessarily allow you to do more but it simplifies formulas which have multiple steps and many different variables.

Here is a very simple to follow shell script that includes bc and creates its own function to be used by bc.

    bcsample.sh

```bash
#!/bin/bash

# bcsample.sh
if [ $# != 1 ]
  then
    echo "usage: "$0" <value>"
    exit
fi

bc <<END-OF-INPUT
scale=2

/* define the function */
define mysquareit(x){
  return(x * x);
}

/* then use the function to do the calculation*/
x=$1
"Processing ";x;" value squared is ";mysquareit(x)
quit
END-OF-INPUT

echo "(to 2 decimal places)"
```

The part of this script that is perhaps unique is the "END-OF-INPUT". It is possible for a shell script to embed text or commands into itself and redirect that information into a program.

This example script does not demonstrate an amazing self defined function but rather that this function is read from the shell script itself.

date - system date and time

The system date and time is important. It is helpful to see when a file was last modified but the date or time can be important in other situations as well. One example of this would be if creating log files or data files. Sometimes it is more convenient that the data files are segregated by date.

The date command will display date, time and timezone.

$ date
Sun Oct 20 17:53:04 CEST 2019

The standard output of the date command does indeed show all information but it is not as convenient for computer programs. Sorting data is easier if it is in the ISO format.

    YYYYMMDD
    HHMMSS

The Linux date command can be used to generate the date or time in these formats so it can be more easily used by programs or scripts.

Common options
+'format'   Desired format sequences

                                   %Y   4 digit year
                                   %y   2 digit year
                                   %m   2 digit month
                                   %d   2 digit day
                                   %H   2 digit hour
                                   %M   2 digit minute
                                   %S   2 digit second

--date=YYYYMMDD   date to use

It is possible to use the date command to set the system time as well but this tends to be a rather seldom event due to the time servers on the internet.

$ echo `date +'%Y-%m-%d %H:%M:%S'`
2019-10-20 18:21:15

Just like most text based programs the output can be easily assigned to a variable and then used for other purposes.

grep - searching for strings

There are instances where you want to find certain patterns in files. This might be looking for errors or warnings.

The grep command like most Linux command line programs can be used either for searching through files or even through standard input when output is piped through this command.

Common options
-i   Ignore case when matching
-n   Also display the line number of the match
-l   Do a file listing of the filenames of the files that match

-v   Invert match. Show things that do not match the pattern

--date=YYYYMMDD   date to use

grep [OPTIONS] <pattern> <filename or regular expression>

grep -i error logfile.txt
grep -in error logfile.txt

find - locating files

Disk drives are now multi terabyte in size and so it can be difficult to remember where a file was stored. Linux has the find command which can be used to locate not only a file of a specific filename it is possible to find files with a certain pattern (e.g extension), files that have been recently modified as well as files owned by a user or group.

Common options
    -L   Follow symbolic links
    -group <groupname>   Files owned by group groupname.
    -user <username>   Files owned by user username.
    -exec <cmd> {} ";"   Execute command cmd on the file found
    -name <name pattern>   Regular expression of filename(s)
    -ls   Directory listing of file that was found
    -inum <n>   This will find the file with the given inode number.

   find <search directory> [OPTIONS]

The find command is one of the most flexible commands. It can be used to find a file based on many different criteria but it is also possible to execute a command or script.

Find all text files starting in the current working directory and going recursively through all sub directories.

   find . -type f -exec file '{}' \;

There are cases where the find statement is the only way to perform a special task. It is possible to accidentally create a file that contains an unprintable character in the name. If you cannot see the character how can you open the file, rename the file or delete the file?

If this happens you can see the inode numbers of each file with the ls command.

   $ ls -il
   4198773 -rwxrwxrwx 1 dock dock 10 Oct 20 19:07 script1.sh
   4198772 -rwxr-xr-x 1 dock dock 10 Oct 20 15:20 script2.sh
   4198774 -rwxr-xr-x 1 dock dock 10 Oct 20 15:20 script3.sh
   4198775 -rwxr-xr-x 1 dock dock 10 Oct 20 15:20 script4.sh
   4198779 -rw-r--r-- 1 dock dock 4 Oct 20 19:11 ''$'\b''ssls'

It is not possible to know what exactly the control character is but we can see that the inode number is 4198779. If we want to rename this file we can use this information along with the find command.

   find . -inum 4198779 -exec mv {} newname.sh ";"

sed - stream editor

The stream editor program is a very unique program even for Linux. It is not unique because it works on streams or files but because it is a "S"tream "ED"itor. It is not very often that you edit files in a non-interactive manner.

The stream editor program sed will allow you to create sets of instructions that can be used to edit a stream or file.

   Common options
       -f   File with sed commands
       -e <sed command>   Sed command to be used against input
       -i   Replace file in place

    sed [OPTIONS] <filename>
    sed [OPTIONS]

A very visual way to testing out sed would be to try it on a directory listing. The output is very well defined and obvious when it makes it changes.

Step 1 - Sample file listing

   $ ls -l
   total 20
   -rwxrwxrwx 1 dock dock 10 Oct 20 19:07 script1.Sh
   -rwxr-xr-x   1 dock dock 10 Oct 20 15:20 script2.Sh
   -rwxr-xr-x   1 dock dock 10 Oct 20 15:20 script3.Sh
   -rwxr-xr-x   1 dock dock 10 Oct 20 15:20 script4.Sh

Step 2 - Converting lower case i to capitalized I.

   $ ls -l | sed -e 's/i/I/'g
   total 20
   -rwxrwxrwx 1 dock dock 10 Oct 20 19:07 scIpt1.sh
   -rwxr-xr-x   1 dock dock 10 Oct 20 15:20 scIpt2.sh
   -rwxr-xr-x   1 dock dock 10 Oct 20 15:20 scIpt3.sh
   -rwxr-xr-x   1 dock dock 10 Oct 20 15:20 scIpt4.sh

Step 3 - Removing total line

   $ ls -l | sed -e 's/i/I/'g | sed /total/d
   -rwxrwxrwx 1 dock dock 10 Oct 20 19:07 scIpt1.sh
   -rwxr-xr-x   1 dock dock 10 Oct 20 15:20 scIpt2.sh
   -rwxr-xr-x   1 dock dock 10 Oct 20 15:20 scIpt3.sh
   -rwxr-xr-x   1 dock dock 10 Oct 20 15:20 scIpt4.sh

Step 4 - Removing everything except for the filename.

   $ ls -l | sed -e 's/i/I/'g | sed /newname/d | sed /total/d | sed 's/^.* //'
   scIpt1.sh
   scIpt2.sh
   scIpt3.sh
   scIpt4.sh

With these few examples we can see that it is possible to change how the directory output looks. This is a pretty contrived example, if we needed to see the only the file names we would use the -1 option for ls.

Once a set of commands have been worked out and proven themselves they can be put into a text file. This list of proven commands can then be used over and over quite easily.

   commands.sed
   s/i/I/g
   /newname/d
   /total/d
   s/^.* //

Additionally the command itself is much easier to look at and thus cleaner as far as future maintenance is concerned.

$ ls -l | sed -f commands.sed
commands.sed
scIpt1.sh
scIpt2.sh
scIpt3.sh
scIpt4.sh

APPENDIX SMD SIZE CHART

Soldering with through hole components is really pretty easy, however, at some point you might want to try your hand at surface mount parts ... or the components you wish to use only come in that format.

Here is a small chart showing the smd size and how that relates to actual human sizes.

Size	Length(in mm)	Width(in mm)
0201	0.60	0.30
0402	1.0	0.50
0603	1.60	0.80
0805	2.0	1.25
1206	3.20	1.60
1210	3.20	2.60
1217	3.0	4.20
2010	5.0	2.60
2020	5.08	5.08
2045	5.0	11.50
2512	6.3	3.10

APPENDIX ARDUINO CUBE PATTERN CODE

```
int column1 = 54;   // analog pin 0
int column2 = 55;   // analog pin 1
int column3 = 56;   // analog pin 2
int column4 = 57;   // analog pin 3
int column5 = 58;   // analog pin 4
int column6 = 59;   // analog pin 5
int column7 = 62;   // analog pin 6
int column8 = 61;   // analog pin 7
int column9 = 60;   // analog pin 8

int POWERON = 1;
int POWEROFF = 0;

int bottomLayer = 41;   // digital 41
int middleLayer = 43;   // digital 43
int topLayer    = 45;   // digital 45

int columns[] = {-1, column1,column2,column3,
    column4,column5,column6,
    column7,column8,column9 };

void groundLayer(int v)
{
  digitalWrite(v,LOW);
}
void unGroundLayer(int v)
{
  digitalWrite(v,HIGH);
}

void columnOn(int v)
{
  digitalWrite(v,HIGH);
}

void columnOff(int v)
{
  digitalWrite(v,LOW);
}

void layersOff()
{
  digitalWrite(topLayer,HIGH);
  digitalWrite(middleLayer,HIGH);
  digitalWrite(bottomLayer,HIGH);
}

void columnsOff()
{
  int idx;
```

```
  for (idx = 1; idx <= 9; idx++)
    digitalWrite(columns[idx],POWEROFF);
}

void columnsOn()
{
  int idx;
  for (idx = 1; idx <= 9; idx++)
    digitalWrite(columns[idx],POWERON);
}

void allOff()
{
  layersOff();
  columnsOff();
}

void setup()
{
  int idx;
  Serial.begin(115200);
  Serial.println("arduino atmega 2560 3x3x3 ");

  // setup all columns for output
  for (idx = 1; idx <= 9; idx++)
    pinMode(columns[idx],OUTPUT);

  // setup all layers for output (e.g as power sink)
  pinMode(bottomLayer,OUTPUT);
  pinMode(middleLayer,OUTPUT);
  pinMode(topLayer,OUTPUT);
}

void floatingLayers()
{
  // set up all columns to use the current power level
  int idx;
  for (idx = 1; idx <= 9; idx++)
    digitalWrite(columns[idx],POWERON);

  for (int iter = 0; iter < 5; iter++)
  {
    Serial.println(iter);

    // turn off all, except for bottom layer of cube
    layersOff();
    digitalWrite(bottomLayer,LOW);
    delay(1000);

    // turn off all, except for middle layer of cube
    layersOff();
    digitalWrite(middleLayer,LOW);
    delay(1000);

    // turn off all, except for top layer of cube
    layersOff();
    digitalWrite(topLayer,LOW);
    delay(1000);

    // turn off all, except for middle layer of cube
    layersOff();
    digitalWrite(middleLayer,LOW);
    delay(1000);
  }
}

void leftDiamond()
{
  for (int iter = 1; iter < 6; iter++)
  {
```

```
    int pause = iter * 100;
    allOff();
    digitalWrite(column1,POWERON);
    digitalWrite(bottomLayer,LOW);
    delay(pause);

    allOff();
    digitalWrite(column3,POWERON);
    digitalWrite(column5,POWERON);
    digitalWrite(column7,POWERON);
    digitalWrite(middleLayer,LOW);
    delay(pause);

    allOff();
    digitalWrite(column9,POWERON);
    digitalWrite(topLayer,LOW);
    delay(pause);

    allOff();
    digitalWrite(column3,POWERON);
    digitalWrite(column5,POWERON);
    digitalWrite(column7,POWERON);
    digitalWrite(middleLayer,LOW);
    delay(pause);

    allOff();
    digitalWrite(column1,POWERON);
    digitalWrite(bottomLayer,LOW);
    delay(pause);
  }
  allOff();
}

void rightDiamond()
{
  for (int iter = 1; iter < 6; iter++)
  {
    int pause = iter * 100;

    allOff();
    digitalWrite(column7,POWERON);
    digitalWrite(bottomLayer,LOW);
    delay(pause);

    allOff();
    digitalWrite(column1,POWERON);
    digitalWrite(column5,POWERON);
    digitalWrite(column9,POWERON);
    digitalWrite(middleLayer,LOW);
    delay(pause);

    allOff();
    digitalWrite(column3,POWERON);
    digitalWrite(topLayer,LOW);
    delay(pause);

    allOff();
    digitalWrite(column1,POWERON);
    digitalWrite(column5,POWERON);
    digitalWrite(column9,POWERON);
    digitalWrite(middleLayer,LOW);
    delay(pause);

    allOff();
    digitalWrite(column7,POWERON);
    digitalWrite(bottomLayer,LOW);
    delay(pause);
  }
```

```
  allOff();
 }

/*
 * lots of pieces for next pattern
 */
void XAxisRotate_1() // horizontal
{
  int pause = 2;

  for (int idx = 0; idx < 5; idx++)
  {
    columnOn(column3);  groundLayer(middleLayer);
    delay(pause);
    columnOff(column3); unGroundLayer(middleLayer);

    columnOn(column6);  groundLayer(middleLayer);
    delay(pause);
    columnOff(column6); unGroundLayer(middleLayer);

    columnOn(column9);  groundLayer(middleLayer);
    delay(pause);
    columnOff(column9); unGroundLayer(middleLayer);

    columnOn(column2);  groundLayer(middleLayer);
    delay(pause);
    columnOff(column2); unGroundLayer(middleLayer);

    columnOn(column5);  groundLayer(middleLayer);
    delay(pause);
    columnOff(column5); unGroundLayer(middleLayer);

    columnOn(column8);  groundLayer(middleLayer);
    delay(pause);
    columnOff(column8); unGroundLayer(middleLayer);

    columnOn(column1);  groundLayer(middleLayer);
    delay(pause);
    columnOff(column1); unGroundLayer(middleLayer);

    columnOn(column4);  groundLayer(middleLayer);
    delay(pause);
    columnOff(column4); unGroundLayer(middleLayer);

    columnOn(column7);  groundLayer(middleLayer);
    delay(pause);
    columnOff(column7); unGroundLayer(middleLayer);
  }
}

void XAxisRotate_2() // uphill
{
  int pause = 2;
  for (int idx = 0; idx < 5; idx++)
  {
    columnOn(column3);  groundLayer(topLayer);
    delay(pause);
    columnOff(column3); unGroundLayer(topLayer);

    columnOn(column6);  groundLayer(topLayer);
    delay(pause);
    columnOff(column6); unGroundLayer(topLayer);

    columnOn(column9);  groundLayer(topLayer);
    delay(pause);
    columnOff(column9); unGroundLayer(topLayer);

    columnOn(column2);  groundLayer(middleLayer);
    delay(pause);
    columnOff(column2); unGroundLayer(middleLayer);
```

```
    columnOn(column5); groundLayer(middleLayer);
    delay(pause);
    columnOff(column5); unGroundLayer(middleLayer);

    columnOn(column8); groundLayer(middleLayer);
    delay(pause);
    columnOff(column8); unGroundLayer(middleLayer);

    columnOn(column1); groundLayer(bottomLayer);
    delay(pause);
    columnOff(column1); unGroundLayer(bottomLayer);

    columnOn(column4); groundLayer(bottomLayer);
    delay(pause);
    columnOff(column4); unGroundLayer(bottomLayer);

    columnOn(column7); groundLayer(bottomLayer);
    delay(pause);
    columnOff(column7); unGroundLayer(bottomLayer);
  }
}

void XAxisRotate_3() // verticle
{
  int pause = 2;
  for (int idx = 0; idx < 5; idx++)
  {
    columnOn(column2); groundLayer(topLayer);
    delay(pause);
    columnOff(column2); unGroundLayer(topLayer);

    columnOn(column5); groundLayer(topLayer);
    delay(pause);
    columnOff(column5); unGroundLayer(topLayer);

    columnOn(column8); groundLayer(topLayer);
    delay(pause);
    columnOff(column8); unGroundLayer(topLayer);

    columnOn(column2); groundLayer(middleLayer);
    delay(pause);
    columnOff(column2); unGroundLayer(middleLayer);

    columnOn(column5); groundLayer(middleLayer);
    delay(pause);
    columnOff(column5); unGroundLayer(middleLayer);

    columnOn(column8); groundLayer(middleLayer);
    delay(pause);
    columnOff(column8); unGroundLayer(middleLayer);

    columnOn(column2); groundLayer(bottomLayer);
    delay(pause);
    columnOff(column2); unGroundLayer(bottomLayer);

    columnOn(column5); groundLayer(bottomLayer);
    delay(pause);
    columnOff(column5); unGroundLayer(bottomLayer);

    columnOn(column8); groundLayer(bottomLayer);
    delay(pause);
    columnOff(column8); unGroundLayer(bottomLayer);
  }
}

void XAxisRotate_4() // downhill
{
  int pause = 2;
  for (int idx = 0; idx < 5; idx++)
```

```
  {
    columnOn(column1); groundLayer(topLayer);
    delay(pause);
    columnOff(column1); unGroundLayer(topLayer);

    columnOn(column4); groundLayer(topLayer);
    delay(pause);
    columnOff(column4); unGroundLayer(topLayer);

    columnOn(column7); groundLayer(topLayer);
    delay(pause);
    columnOff(column7); unGroundLayer(topLayer);

    columnOn(column2); groundLayer(middleLayer);
    delay(pause);
    columnOff(column2); unGroundLayer(middleLayer);

    columnOn(column5); groundLayer(middleLayer);
    delay(pause);
    columnOff(column5); unGroundLayer(middleLayer);

    columnOn(column8); groundLayer(middleLayer);
    delay(pause);
    columnOff(column8); unGroundLayer(middleLayer);

    columnOn(column3); groundLayer(bottomLayer);
    delay(pause);
    columnOff(column3); unGroundLayer(bottomLayer);

    columnOn(column6); groundLayer(bottomLayer);
    delay(pause);
    columnOff(column6); unGroundLayer(bottomLayer);

    columnOn(column9); groundLayer(bottomLayer);
    delay(pause);
    columnOff(column9); unGroundLayer(bottomLayer);
  }
}

void XAxisRotate()
{
  for (int idx = 0; idx < 5; idx++)
  {
    XAxisRotate_1(); // horizontal
    XAxisRotate_2(); // uphill
    XAxisRotate_3(); // vertical
    XAxisRotate_4(); // downhill

    XAxisRotate_1(); // horizontal
    XAxisRotate_2(); // uphill
    XAxisRotate_3(); // vertical
    XAxisRotate_4(); // downhill

    XAxisRotate_1(); // horizontal
    XAxisRotate_2(); // uphill
    XAxisRotate_3(); // vertical
    XAxisRotate_4(); // downhill

    XAxisRotate_1(); // horizontal
    XAxisRotate_2(); // uphill
    XAxisRotate_3(); // vertical
    XAxisRotate_4(); // downhill

    XAxisRotate_1(); // horizontal
    XAxisRotate_2(); // uphill
    XAxisRotate_3(); // vertical
    XAxisRotate_4(); // downhill

    XAxisRotate_1(); // horizontal
    XAxisRotate_2(); // uphill
```

```
    XAxisRotate_3();  // vertical
    XAxisRotate_4();  // downhill
  }
}

/*
 * lots of pieces for next pattern   (z axis)
 */
void chase()
{
  int pause = 300;
  int levels[] = { bottomLayer,middleLayer,topLayer,middleLayer};

  for (int iter = 1; iter < 6; iter++)
  {
    pause = iter * 50;
    for (int x = 0; x < sizeof(levels) / sizeof(levels[0]); x++)
    {
      allOff();
      columnOn(column1);  groundLayer(levels[x]);
      delay(pause);
      columnOff(column1); unGroundLayer(levels[x]);

      columnOn(column2);  groundLayer(levels[x]);
      delay(pause);
      columnOff(column2); unGroundLayer(levels[x]);

      columnOn(column3);  groundLayer(levels[x]);
      delay(pause);
      columnOff(column3); unGroundLayer(levels[x]);

      columnOn(column4);  groundLayer(levels[x]);
      delay(pause);
      columnOff(column4); unGroundLayer(levels[x]);

      columnOn(column5);  groundLayer(levels[x]);
      delay(pause);
      columnOff(column5); unGroundLayer(levels[x]);

      columnOn(column6);  groundLayer(levels[x]);
      delay(pause);
      columnOff(column6); unGroundLayer(levels[x]);

      columnOn(column7);  groundLayer(levels[x]);
      delay(pause);
      columnOff(column7); unGroundLayer(levels[x]);

      columnOn(column8);  groundLayer(levels[x]);
      delay(pause);
      columnOff(column8); unGroundLayer(levels[x]);

      columnOn(column9);  groundLayer(levels[x]);
      delay(pause);
      columnOff(column9); unGroundLayer(levels[x]);
    }
  }
}
void loop()
{
   chase();
   floatingLayers();
   leftDiamond();
   rightDiamond();
   XAxisRotate();
}
```

GLOSSARY

0.1" The standards for PCB using through hole components are usually with the spacing of 0.1 inch (0.254 cm). This American measuring standard is less used in surface mount components which are usually measured in millimeters.

Active component A device that can actively manipulate the signal. One example of a active component is a micro-controller.

AG The symbol for silver from the periodic table of elements.

Artwork The design for printed circuit board.

AVR Micro-controller family developed by Atmel.

AVR Programmer This is a small board that is used in conjunction with software to program an Atmel micro-controllers from your personal computer.

Bit mapped bytes All bytes are composed of 8 bits. Bit mapped memory is when each bit is treated by the program as an array of boolean values.

BI The symbol for bismuth from the periodic table of elements.

BGA Ball grid array package format.

Cold joint A cold joint is where the solder does properly melt and join all the components together. The characteristics of a cold solder joint is that they are lumpy and usually do not have a clean shine. Over time cold solder joints can crack and will be unreliable over time.

CAD Computer aided design. Here, this refers to the software used in the designing of the schematics and pcb layout.

Component This can be any individual active or passive element that uses or manipulates the signal. This may include battery or power connectors.

Component side The through hole components are usually only placed from one

	side of the printed circuit board.
CU	The symbol for copper from the periodic table of elements.
Current	This is the volume of the electron flow. The units of measurement for current is in Amps.
Design rules	This is usually a list of guidelines or tolerances that must be met in the printed circuit board layout. (e.g. traces must be a certain distance from edge of board, minimum or maximum sizes for holes)
DIP/DIL	Dual inline package. This is one of many different formats that transistors chips may be.
DRC	Design rule check. This is a process of the CAD software which checks to ensure that none of the design rules have been violated.
Drill file	This file contains the coordinates and sizes of the holes to be drilled. The format of the file is usually the industry standard "Excellon drill file format".
EPROM	A read only memory that can only be erased by ultraviolet light.
EEPROM	Electrically erasable programmable read-only memory. This is a type of non-volatile memory used in devices to store small amounts of data. This data can be erased and reprogrammed in individual bytes.
Eutectic	Eutectic mixture is one that melts or freezes at a temperature lower than any of the components of that mixuture. Solder is an eutectic mixture, usually composed of tin and lead.
Footprint	The pattern or arrangement of the pads (SMD) or the legs of a component used to attach the part to the printed circuit board.
Fuse bytes	Configuration bytes which determine which functionality will be enabled in the Atmel chip.
Gerber file	Gerber files are a defacto industry standard format for describing printed circuit boards.
Hertz	The hertz is the derived unit of frequency defined as one cycle per second.
I2C	Inter-Integrated circuit which is usually referred to as I2C or I2C. This protocol was created by Philips Semiconductor in 1982 to connect peripheral chips in televisions. See also TWI.
IDE	Integrated development environment. These environments usually support the following functions editing, compiling, and debugging.
ISP	This is an in-circuit serial programmer which allows your computer to transfer programs to some programmable devices such as micro-

controllers. Some PCB's such as the Arduino can be connected up to an ISP and can program the micro-controller even while installed in a complete system.

Lead
: An electrical connection consisting of a length of metal wire or a metal pad designed to connect two locations electronically.

Make
: A command line program that given a list of dependencies will check to see if any source files need to be generated. Normally this is compiling or linking but can be any command line program which may be needed during development.

Makefile
: The make command reads it configuration file "Makefile" from the current directory. This file contains the dependencies as well as the rules and commands for generating objects.

mF
: Microfarad. This is a unit of capacitance, equivalent to 0.000001 (10 to the -6th power) farad. Due to the symbol mu(m) being less common for fonts, it is often substituted with uF which is in all fonts.

Micro-controller
: A compact integrated circuit that contains a processor, memory and input/output pins to support general I/O operations. Micro-controllers tend to be small with limited resources and are used in embedded applications.

Mil
: The definition of 1 Mil is 1/1000 of an inch. This measurement scale is typically used when describing the minimum required spacing between traces, holes, or other objects on the PCB. One Mil is not one millimeters but rather 0.0254mm.

Nets / Net list
: A network is simply a trace that has been given a name. This trace may or may not be directly connected to all portions of the schematic. Instead of the trace connecting all related components it is common that in one or more components simply has a tag or label with the network name. Thus each label makes up a connection to the virtual network with that tag name. Just like a good program, nets are giving a meaningful name based ont he function of the signals that go over that trace. A power network may be called "VCC", "power" or "5V".

Passive component
: A device that may manipulate the signal but behaves in the same manner. An example of a passive component is the resistor or capacitor.

PB
: The symbol for lead from the periodic table of elements.

PCB
: A printed circuit board is a thin board usually made of fiberglass or some other laminate material. The copper paths that make up the circuit are etched into the board. These paths connect and the board mechanically supports electronic parts.

PWM	Pulse width modulation is a modulation technique which is used to vary the amount of power being supplied to another electrical device such as a LED or a motor.
Ratsnest	The connections between the components in the PCB editor are displayed with a thin line. This can suggest where the traces can be connected.
Resistance	Resistance is the prevention of electrons flowing. Resistance is measured in Ohms.
Routing	The design of the individual of traces on a printed circuit board.
Schematic	The abstract representation of the electronic circuit.
Silk screen	The letters or other printing which is used to identify the components, part numbers, symbols or logos that is printed on the PCB. The name derives from the process of how the ink is actually applied to the board. The color is usually white but is not required to be white.
Sketch	A sketch is the name that Arduino uses for a program.
Solder mask	Is a thin lacquer like layer that is placed around the copper traces to prevent solder bridges from forming when soldering closely spaced solder pads.
SMD	Surface mount device. These types of parts are soldered directly to the PCB board. Exactly as the name implies the part is soldered directly to one side of the PCB with no through holes.
SN	The symbol for tin from the periodic table of elements.
SoC	System on a chip. A system on a chip (SoC) combines various electronic circuits and components into a single, integrated chip (IC). The SoC may contain support for PWM, RAM, cpu or even support for graphical processing. All of this technology is contained in a single package.
SPI	Serial peripheral interface is a bus for transferring data between micro-controllers, sensors, or other devices.
Tarball	A tarball file is a file that has been created by the tar command. The tar command will create a archive file that can contain files and directories. This is quite similar to the Zip file that is typically seen on Windows. Tarball files are a convenient way to transport source code for a application from place to place.
Tinning	The act of applying a small amount of fresh solder to parts that will be later joined together. The flux in the solder will remove oxidation

	and oils and make it easier to join the parts in a subsequent step.
Through hole	Electronic components that has legs that are designed to go through the PCB to be soldered.
TO-92	A three legged semiconductor package used mainly for transistors.
TQFP	Thin quad flat pack package format.
Track / Trace	The copper lines on a printed circuit board that allows the electricity to flow between the electronic components.
TWI	Two wire interface, TWI, which is supported on the Atmel family of micro-controllers. This TWI actually has been developed to be compatible with I2C from Philips Semiconductor. See I2C.
UART	Universal asynchronous receiver/transmitter is the microchip that allows serial transfer of data programming to the attached device.
Via	A via is a connection between two pads which each exist on different layers of the printed circuit board. This hole is electroplated so that these pads are electronically connected between the two layers at this spot. A via is used when a signal needs to be passed between the two layers.
Voltage	This is the difference of the potential between two points. The units of measurement for voltage is Volts.
ZIF Socket	Zero insertion force socket. This type of socket requires very little force to insert socket into and usually has a mechanism which will lock down the chip so it cannot fall out.

ABOUT THE AUTHOR

Christopher was born in the United States in the Midwest and started his career as a software developer but apparently became restless and made the move to Europe. He has spent most of his time working in IT departments for banks and large corporations, usually as a consultant but occasionally as a software developer. His work brought him in contact with different people from different cultures but typically in a workplace environment.

It was during one of the IT projects he was lucky enough to meet an old time electrical engineer from the former Soviet Union. During the daytime we worked on software and in the evenings he infused me with the joy of hardware development.

www.ingramcontent.com/pod-product-compliance
Lightning Source LLC
Chambersburg PA
CBHW080634230426
43663CB00016B/2863